DIRT HOG

A HANDS-ON GUIDE TO RAISING PIGS OUTDOORS . . . NATURALLY

DIRT HOG

A HANDS-ON GUIDE TO RAISING PIGS OUTDOORS . . . NATURALLY

KELLY KLOBER

Acres U.S.A.
Austin, Texas

DIRT HOG

A HANDS-ON GUIDE TO RAISING PIGS OUTDOORS . . . NATURALLY

Acres U.S.A.
P.O. Box 91299
Austin, Texas 78709 U.S.A.
(512) 892-4400 • fax (512) 892-4448
info@acresusa.com • www.acresusa.com

Printed in the United States of America

Publisher's Cataloging-in-Publication

Klober, Kelly, 1949-
Dirt hog / Kelly Klober. Austin, TX, ACRES U.S.A., 2007
vi, 309 pp., 23 cm.
Includes Index
Includes Bibliography
Incudes Illustrations
ISBN 978-1-60173-001-5 (trade)

1. Animal husbandry — Swine. 2. Swine — Feeding and feeds.
3. Swine breeds. 4. Swine diseases. 5. Organic farming.
6. Swine folklore. I. Klober, Kelly, 1949- II. Title.

SF395.K54 636.4

CONTENTS

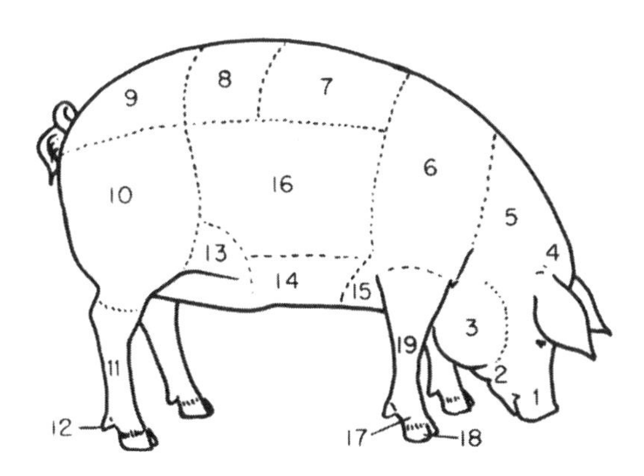

CHAPTER 1

THE RETURN OF THE RANGE HOG

When the Spanish came to the New World in the 1500s, the animal they brought with them that drew the most acclaim was the horse. It was an animal returning to the lands of its prehistoric origins. The more important creature they carried, however, may have been from the family Suidae (*a.k.a.* "sui")—the hog.

For the Conquistadores, it was a commissary on four good legs, able to move with them as they charted new lands. Descendants of the wild boars of Europe, these creatures could range out under the eyes of a watchful swineherd to largely fend for themselves. For centuries, hogs essentially sustained themselves on what could be gleaned from roughly one foot below to one foot above ground level. They were sometimes the sanitation service in Colonial America and trailed the settlers West with the expanding frontier. There hogs lived a free-ranging existence along the early streets and byways. A part of their subsistence was discarded table wastes, sometimes tossed unceremoniously from the back doorstep.

Hogs were and are omnivores; they will eat fallen fruits and nuts, insects, eggs, and just about anything bite-sized that they encounter in that two foot zone. They still have a valuable role to play in gleaning the residues of grain harvests and are fair grazers, unaffected by bloat from an all-legume pasture mix.

The hog was probably the animal for which the concept of "free range" was developed. Early on in the New World, they were set loose in the woodland margins around settlements and homesteads and left alone to meet their own needs. "Root hog or die" pretty much described life for both Man and beast along the early frontier. There they would breed and reproduce seasonally, meeting nearly all of their own nutritional needs (grain and scraps might be fed to a hog or two penned up for a short time prior to slaughter), and early fencing was actually used to keep hogs and other ranging livestock out rather than in. It wasn't livestock breeding, it wasn't even really livestock keeping, but it was an early expression of what the ranging hog could do and become.

Before there were cattle drives, there were hog and poultry drives up in the Northeast. In Colonial America, hogs were a livestock variety hardy enough to cope with harsh winters and the cold and damp of the spring and fall. They could be rounded up and moved up and down the pikes and trails of old in what were often quite large droves. Many parts of the Northeast and New England still maintain a claim as producers of truly premium cured pork products. Many sugar cures were born there.

The hardiness and the vigor of the hog enabled the animal to move freely with the American frontier. Across the Cumberland Gap, through the "dark and bloody" of ol' Kaintuck, and out onto the prairies they went. They were the porcine equivalent of Cooper's leather-clad frontiersmen. They were the early economic base for such great Midwestern cities as Cincinnati and Chicago and spawned many of the great fortunes of the 19th century. Is there a hog producer anywhere who cannot acknowledge truth in words of poet Carl Sandburg, who called Chicago "hog butcher to the world"?

Hogs were thus builders of both fortunes and empires. And to this very day, they have the fastest turnaround on investment among the large animal species, as well as the shortest time to income.

The grain-producing potential of the heartland was recognized early on, but it was not easily transported and marketed using the technology of the day. Other than in a narrow wagon bed or a few hard-to-handle sacks, the only way to bring coarse grain to town was as "white whiskey," or cured pork. Cured pork pieces could be swung over a saddle horn and often later served as actual units of currency along the frontier.

The value placed on good hogs in that earlier day is well illustrated in the following anecdote. An early breeder of Berkshire hogs in Missouri owned a house in the early French style, with roofed porches on all sides of the house. At nightfall, he would bring in his valued charges to sleep safely confined on those very porches.

Being from Missouri, I must share at least one story from Mark Twain about hogs, a species that he greatly admired. In Mr. Twain's youth, hogs ranged freely throughout the Show-Me state. Early rural churches then had puncheon floors: split logs laid exposed-side up and through which cooling drafts would flow in warm weather. In that early day, church doors were left open to provide refuge for travelers, and on very hot days, hogs would enter the churches to lie upon the cool floors. This prompted Mr. Twain to write, ". . . the principal difference between the men and hogs of Marion County was that the hogs wanted to go to Church."

From coast to coast and even to the Hawaiian Islands, hogs were brought and found a home in all of our 50 states. The actual date of the first domestication of the hog is lost in time, but many believe that it may date as far back as 6,000 years ago. Six thousand years to at last arrive in the U.S. heartland to find the niche in which they would at last fully and most profitably bloom.

Foreign breeds like the Berkshire and Hampshire were refined and honed here, while the Duroc and the Chester White breeds were totally American creations. They provided the four good legs on which much of the grain wealth of the hinterland could be driven to consumers awaiting in America's cities.

Early in the last century, farmers were endeavoring to create a swine-producing system that would assure both animal health and farm productivity. This at a time when there were very few drugs and there was no such thing as an environmental movement. They recognized early on that concentrating hog numbers (or any livestock, really) in one small area would have very extreme consequences. The nature of the hog's pointed hoof would take a severe toll on earthen surfaces, and they realized that rotating pasturage was every bit as beneficial as rotating cropping ground.

Taking its name from McLean County, Illinois, where the system was refined, the McLean County System of swine production is still as close to perfect as swine-raising has ever been. Essentially, it boils down to four steps:

1. Farrowing is done in a central area that is regularly cleaned and disinfected between farrowings. There is no continuous farrowing cycle and thus no perpetual pool of very young and lactating animals, which are often the epicenter of health problem outbreaks.

2. As soon as possible following farrowing, the sows and litters are moved to pasture or wooded lots. This is done on a trailer or other conveyance that is thoroughly cleaned and disinfected between uses.

3. Each group remains intact through to weaning, and this is done by removing the sows and leaving the pigs in a familiar environment to reduce stress. A later modification was to remove the pigs to simple, open-fronted nurseries that have been cleaned, disinfected, and allowed to stand idle for some time between pig crops.

4. Pastures or lots are then kept free of hogs for at least 12 months to break the life cycles of most pathogenic organisms and parasites that might be harbored in the soil. Often, such lots and pastures would go back into the crop rotation and not see hogs again for as many as 3 or 4 years.

A succinct definition of this venerable system was a clean start, a clean ride, and clean ground. "Clean" here meaning free of the prolonged presence of hogs, their wastes and fecal material, and the harmful organisms they might harbor. The year-long break between uses is believed to be sufficient to break the lifecycles of most disease organisms and parasites.

A shortcoming was never found in the McLean System, and glimpses of it can still be seen in pasture-utilizing systems in western Illinois, southeastern Iowa, and eastern Missouri. Burgeoning land prices took a greater toll on the system than anything else. I can never buy into the argument that farmland is too valuable for livestock production, since it is their worth as feedstuffs that still give corn and soybeans their greatest value. New generations of wormers and external parasite control products have likewise come along to facilitate a more intensive land-use pattern.

Some aspects of this system may at first appear quite daunting to the producer with a small land base, but it shouldn't be. In those situations, a rest of as few as 90 days between uses can help break health problem cycles. Also, with smaller numbers, the stress load is greatly reduced, and modern parasite controls are available in large enough numbers to regularly rotate their use to prevent resistance buildup. Perpetual farrowings and the crowding of animals through available spaces and facilities can all too quickly have a snowball effect on health problems and consequences that the producer with modest numbers might never encounter.

Range Hardiness Comes Naturally

Contrary to the impression given by the levels of biosecurity now practiced in some confinement units, the hog is not a health disaster just waiting to happen. These are, after all, the creatures that walked boldly across the whole of the American continent and into all sorts of varied climates and environments.

They are bull-horn tough when allowed to function in a natural manner. They can take a lot in stride when not hobbled by the stress and, some would even say, the torment of confinement rearing. In a confined environment, however, even spilled food particles can become stress-increasing abrasives to skin and hoof pads. There is no yield or give in that stark and unrelenting environment.

A friend I know operates both a purebred Hampshire herd and another producing Hampshire X Yorkshire females for medium- and large-sized pork producers across the Midwest. (For ease of reading, the notation "X" indicates a crossbreed.) His breeding stock ranges freely across stream-drained pastures and stubble fields throughout his sizable cropping operation.

Some would say that out in the open and spread about as they are, they are vulnerable to disease organisms from nearly countless sources. Airborne, waterborne, across-fence contact, bird-carried organisms, and more could occur. Is he fearful? Far from it.

On my first visit to his farm, he drove me out to see the best herd of Hampshire sows I've ever seen. He stepped out of his pickup to bring them closer and bid me to follow him. I started to decline and give the standard answer, which is that some harmful organisms might be carried in my clothes and/or shoes.

His answer to my protests will surprise many; it was to get on out there. As far as he was concerned, if I had brought along anything from home that they hadn't seen, it was time to get them exposed to it, to start building some natural immunities to it. Hogs have walked this Earth with us for 6,000 years because we simply let them be hogs, and the partnership can continue for eons more as long as we let them continue in a simpler, more natural state.

Confinement rearing has been about creating an environment not so much for the hogs themselves, but for the comfort and image of the producers of those hogs. We all have tales of opening a one-sow farrowing hut and marveling at the sow and litter comfortably and securely nestled there. We may be standing there with cold rain running down our shirt collar or in calf-deep snow, but the hogs themselves are free to function as fully as possible.

Within a confinement unit, the hog's world may be limited to a few square feet of concrete, slats, or expanded metal, with a view of dusty walls and battered gates that never change. When the boar test stations were still in place a few years ago, it was a common practice to back up the boars entered there (to be tested on their growth and carcass merits) with a peer more freely housed and less driven back home. This was because such hard-driven hogs were too often socially stunted and unable to do even that most basic of tasks, reproduce.

One school of thought holds that what confinement production actually does is allow mediocre producers to continue to do a mediocre job, but a mediocre job with a great many more hogs. Controlled environments and hyped-up diets have overcome even the beneficial aspects of natural selection, and thus same animals are allowed to remain in the gene pool and breed on, despite their poor genetic potential.

I hate to recall the times I've seen steroids pumped into unsound animals and hormones used to trigger breeding activity. I was once a part of an FFA-type committee that screened hogs ahead of a breeding stock auction. Our veto of an unsound boar reared on a concrete pad was often countered and overridden by a sad but true argument: "Give him a shot and he'll walk out sound on sale day." Concern for the new owner when the shot wore off was simply brushed aside. He may never have bought another purebred boar, became soured on swine testing, or left swine production entirely.

Swine producers that have turned away from purebred genetics or who are soured entirely on swine production are sorely missed now in the struggle against the corporate farms. Given enough hormones, a barrow (a male hog castrated before sexual maturity) will attempt to mount and breed sows. Dead-head and stunted or environmentally damaged boars only do greater damage when allowed to breed on through such artifice.

Why Outdoor Production Now?

What range-rearing of swine can bring almost immediately to the family farm is one more venture for diversification and added cash flow. Perhaps no other large animal enterprise offers as fast a turnaround on an investment.

A sea of ink has been spilled on the subject of risk management for the family farm. And yet the best methodology for it—enterprise diversification and the resulting regular cash flow—were proven long ago. Many years ago, an elderly gentleman set me down and laid out the farming facts of life in quite simple terms.

World Wars and global depressions had shown him much. He distilled his ideas down into a pretty neat rule of thumb for family farm survival. What was needed to endure and succeed were 10 sows, 10 cows, as many ewes, 100 hens, and 100 acres of row crops unencumbered by debt.

Agribusiness has been swept over by more fads over the years than any three generations of teenaged girls. One that has just recently crashed and burned was the idea that a swine producer must be a narrowly focused production specialist; you had to be a farrowing specialist, weaning technologist, or swine finisher extraordinaire. Thus, you end up putting all of your eggs into one narrow basket and watch as the bottom falls out at fairly regular intervals.

On a diversified farm, a modest number of sows and the offspring they produce can do a great number of good things. Among them are (1) they even out cash flow with sales throughout the year, (2) more fully utilize available labor, (3) make it possible to efficiently harvest crop residues, (4) offer a variety of marketing options, (5) serve as are a comfortable fit with a number of other livestock and poultry ventures, (6) have modest start-up and operating costs, and (7) provide a quick turnaround on the investment.

For generations, hogs have had a well-earned reputation as mortgage lifters. They had modest facility needs, produced in good numbers, grew rapidly on a wide variety of readily available feedstuffs, and had ready markets. Within 8 weeks of farrowing you can have lightweight feeder pigs to sell, and within 160 days of birth you can have hogs at a good butcher weight. The females can be bred by 8 months of age to then farrow at just 1 year of age. You can have as many as three litters of pigs on the ground in the time it takes a heifer to produce a single calf, and while twins are considered good performance for a ewe, a gilt (a young female swine) can easily produce four times that number in a single farrowing.

Equipment for keeping hogs can be as simple or as elaborate as your tastes and wallet will permit. Awhile back I saw pictures of hog houses made from discarded heavy equipment tires bolted three high and topped with a few sheets of tin, with a wedge-shaped cut through the bottom two tires for a doorway. The range producer will build using portable equipment and fencing options that are simple to erect and to restructure as lots and pastures are changed or rotated.

The key to success with hogs is the fit, both with the producer and his or her farm and farming mix. The producer has to have an affinity for hogs to succeed with them. There are folks who simply do not like hogs or the thought of tending them. If working with swine is too great a chore, performance will suffer and returns will be diminished. Moreover, they have to fit a home farm that is unlike any other.

Other litmus-test questions to ask are: Are dependable market outlets available? Will the swine enterprise have to compete directly with other ventures for producer time and available resources?

Getting to Know the Range Hog

About a year ago, I was at an event sanctioned by the American Livestock Breeds Conservancy at which a number of rare and endangered livestock breeds were being displayed. One young man was trying most earnestly to get his wife interested in the Red Wattle hogs on display there. He very much wanted to begin raising hogs, and while both professed limited knowledge of the animals, she especially harbored all of the classic doubts and fears. Above all, she was convinced that the knee-high Red shoats (which are young hogs less than 1 year old) before her combined all of the worst elements of the crocodile, pit bull, and Cape buffalo.

The hogs before her were pasture hogs, kid-reared pasture hogs. The greatest threat they posed to anyone was to flop down in the path and refuse to move until they got a good scratching behind the ears. Still, many people, even when only a few generations removed from any type of farming experience, harbor many undue and wrong-headed thoughts about "free ranging" or "wild" hogs.

My grandfather was raised in northern Alabama in the teens and '20s of the last century, and he had a healthy respect for the ridge-runners and razorbacks that roamed the canebrakes and piney woods back then. They were indeed very protective of their young and could make quite a show of tusks and popping jaws.

Those animals, however, had been allowed to revert to a near totally wild state. They ranged "from hell to breakfast" and found and ate what they could. What few shoats that could be rounded up each spring bore the distinctive ear notches of their owners, and built into nearly every barn back then was a "hog trap." It would be baited with grain to lure the near-feral hogs in, and then sliding gates or doors would be tripped behind them. Thus would they be caught and held for a bit of grain finishing or slaughter.

With his family, my Dad traveled to their sharecropped farm in Alabama by covered wagon in 1915. It had been a hard trip that used up most of the family's resources. In their early days there, those free-ranging hogs were almost the entire source of their sustenance. Dad often said that their first meal on their new Alabama home was fried from one end of a hog while his dad was still dressing out the other end.

Indeed, hogs are omnivores, and in times past they had a fearsome reputation for applying the tooth test to just about anything that came their way. They can and still do make a somewhat fearsome display when they feel threatened or pushed. Around hogs or any other large animals, it is always good policy to follow the old stockman's adage of leaving yourself at least two outs when working in close with large stock. Still, modern range production in no way resembles the free-ranging practices of old; domestic hogs have been selectively bred for docility and a gentle, even temperament for generations. Balky sows don't last long on American farms now.

The Tamworth breed, for example, is an old line or heritage breed long known for its hardiness and grit, even in sometimes quite harsh circumstances. A longtime commercial swine producer in the rolling timber along the Missouri River related the following tale about these marvelous animals. He took 10 Tamworth sows with 101 pigs and more or less dumped them into a wooded lot with a self-feeder, creek water to drink, and little else. He returned many days later to find eight of the sows circling the pigs protectively while the other two sows were busy dispatching an intruding coyote. A few weeks later, he weaned all 101 pigs from those brush-running Red sows.

They do sound truly fearsome, don't they? Just let me offer one more true tale about this great old Red breed. A former national president of the Tamworth breed group lived in our county and raised the breed for decades. Later in life, he developed a debilitating disease that often caused him to stumble and fall. He continued raising his beloved Tamworths and would often fall in and around sows with newborn and very small pigs. He never suffered so much as a scratch from these sows, long noted for their vigor and protectiveness.

In an interesting event that stands in stark contrast to this comes from a bit of personal experience. A few years back we bought a couple of gilts from a producer that fed out all of his hogs on a vast concrete feeding floor. We walked the gilts down a long alleyway to one of the most elaborate loading facilities I had ever seen. It was made of heavy-gauge steel pipe and had all sorts of hinged and pivoting gates for crowding and sorting balky hogs.

I commented to the producer on its elaborateness and sturdiness, and his answer was quite telling and reflective of the confinement farming experiences of a great many. He said, "It cost a good bit, but I had to do it. You know they go kinda crazy out there on all that concrete."

Hogs are actually very social animals and are quite safe and easy to handle, as long as you avoid situations that are too forced or overly rushed. Most farmers, for example, will feed the animals over a fence not because of any perceived ferociousness, but because hogs have a tendency to function as a group and curiously crowd around any source of activity near them. When loading or relocating hogs, it is best and simplest to keep them moving forward by blocking behind them.

Make the animals in a lot or pasture aware of your approach by whistling, humming, or talking to them softly. One of my first jobs on the farm was as the officially designated hog caller. I thought it was a sign I was growing up, but my high, youthful voice uttering "whoa, sow" simply carried farther. I was a bipedal "hog whistle" of sorts.

The hog on the range in many ways functions as a free agent. It isn't a wild or uncontrolled animal, but in some respects the hogs do get closer to nature and their animal origins. Domestication and breed development were guided by economic and environmental needs. The hog was thus advanced as a very practical working animal that could produce meat for a broad spectrum of farm types and farming methods. To be sure, it was seen as an ideal consumer of cropping wastes and coarse grains, but it was also a prime range feeder, able to forage and feed on a wide array of feedstuffs.

The other truly great swine myth is that if left to its own devices, the hog is a garbage receptor and a filthy animal. Their hard, sharp hooves can quickly tear up the ground where they congregate, and mud can quickly develop in most areas. This hoof action actually does more lot and pasture damage than any actual rooting activity.

Rooting is part of a hog's natural feeding activity as it forages for roots and grubs just below the soil's surface. This activity can be controlled or even eliminated with a number of different humane devices. Hogs wallow to cool themselves through a process known as evaporative cooling. All such practices are key to maintaining their health and well-being.

In most cases, frequent lot rotation, placing feeders and waterers on simple platforms, and changing feeding sites often effectively counters the problem of mud building up in a particular lot or pasture site. Overstocking a site is also something that should be avoided.

Wallows are also typical of hog habitats. Since these animals also lack the ability to offset the effects of high temperatures by perspiring, the only ways for them to reduce temperature stress is with panting and evaporative cooling; that is, they cool down as the water evaporates from their hides (made longer with mud). So hogs get wet and will instinctively create and/or expand muddy wallows for this purpose. The choice for them is simple: be wet and muddy or hot and dead. If provided with good sources of shade, however, mud development is kept in check.

Through these kinds of management practices, hogs can be kept very clean and pleasant animals to work with. In times past, hogs were set to feed on garbage and little else. In fact, many textbooks can still be found with pictures of lank and spindly hogs rooting through the open sewers of Colonial American towns. Garbage feeding has all but disappeared from the American scene, and the modern hog on range is a grazing and gleaning animal and not a garbage processor.

The offensive image of the hog is actually being propagated anew as they are transported from confinement finishing units. In these circumstances, the animals have frequent contact with their own wastes as they are expelled onto hard and unyielding concrete and steel surfaces. The remaining wastes then often fall through the flooring into large pits directly beneath where the hogs lie. A trailerload of confinement-reared hogs traveling down the road has a truly distinctive odor. It is a smell that has even discouraged a great many veteran hog producers from eating their own production and has probably created more than a fair share of vegetarians.

The hog is also still struggling with a misbegotten image as a fat and wasteful creature. There are still a few hogs out there carrying some excessive cover, but I can truthfully say that I have seen Number 3 and 4 butchers only a handful of times in my 35-plus years with hogs. Most hogs now are very trim and efficient converters of feedstuffs and have been selectively bred to be so for decades.

As an example, nearly 30 years ago, we culled a couple of gilts from the keeper pen and sold one to a family friend for freezer pork. A couple of days later a call came from the processing plant saying that there would be an added cost for working up the gilt. They were going to have to add fat to the trimmings from that gilt to assure pork sausage with good cooking qualities and flavor.

A bit of fat on a hog is a good thing, as it does everything from adding a bit of natural insulation to maintaining good hormonal activity to augmenting the taste of the meat and its cooking qualities. A modern hog will convert protein concentrates to lean gain at a ratio of 4 to 1, and many are now hanging carcasses with a lean yield crowding 70%.

Another porcine myth that needs busting is that the hog is an animal that works best only in the Corn Belt and the South. A certain degree of regionalism does persist with many types of livestock production, but the hog seems to be at home just about anywhere human wandering has taken it. Virginia and New England have been championing the quality of their cured pork products since before the Revolutionary War.

Early in 2003, I received a letter from a young lady in New Hampshire who, with the aid of her father and other family members, had found a strong local market for freezer pork and cured pork product. They were building a largely purebred Duroc herd to approximately 30 sows to meet this strong local demand. Another correspondent in the state of Washington has reported a strong interest in feeder pigs for small family farms there.

A large number of hogs are now being raised in the Western states of Oklahoma, Texas, New Mexico, and Utah. There they are raised for trade in select ethnic markets, as part of youth project work, and for a growing packing industry targeting the still-growing West Coast markets. Hogs are a very popular youth project animal in the state of California as well.

Finally, the hog has had to endure something of a "poor relation in the barnyard" image (as have their keepers). Beef cattle producers have that almost glamorous cowboy image, dairymen are farmers that are often seen as clad head to toe in starch white clothing, while hog farmers wear knee-boots in August and have aroma issues. In reality, hog-rearing can be just as cultured as anything else. In fact, people as distinguished as members of the British royal family keep and promote hogs, including the rare and very endangered English Large Black breed. Prince Charles has been very active in preserving and promoting the Gloucestershire Old Spot, the legendary orchard hog of Great Britain. In France, hogs have gained celebrity for their ability to sniff out those very costly truffles.

The hog has had a profound impact elsewhere, too: at the kitchen table. No other meat animal has so dominated a specific meal as the hog has for the American breakfast. Not only are pork cuts among the most flavorsome meats, they are also very versatile in their uses, and good hogs can produce gourmet cuts that compete with the finest steaks and roasts. And of course it goes without saying that no summer barbecue season would be complete without a cut or two of pork on the grill every weekend.

And then there's the monetary edge hogs bring. The mortgage-raiser reputation was not given to the hog without good cause. Even in a seasonal production pattern, hogs can produce two litters of eight or more apiece each year. These pigs will grow quickly, producing several paydays each year, and will fit a great many different farm enterprise mixes.

How Now Confined Sow?

When production agriculture deigned to pass from simple hog-keeping to a large-scale production of commodity pork, it opted for an industrial model. Its goal became nothing more than to produce as much pork product per worker as possible.

Essentially, a hog's existence was boiled down to the smallest number of square feet possible in a sheet-metal island afloat on a sea of wastes, and it was fed a diet formulated with a computer-generated list of the cheapest possible ingredients. This was designed for whichever community was the most willing to prostitute its tax base and the local environment. It has often and quite rightly been said that the confinement unit of today has been built more to appeal to the eye and the comfort of the operator than to the actual needs of the animals housed there.

The results are a badly compromised end product, a troublesome way to rear hogs that can be fairly challenged on both humane and environmental grounds, and a new kind of serfdom for what were once independent pork producers.

In a sense, the hog really wasn't consulted about being taken indoors and being genetically restructured into something akin to the porcine equivalent of ball bearing: designed to roll effortlessly around and through the corporate machinery. The animal was thus placed into a truly stifling environment, often directly above its own wastes, and then driven as hard and fast as possible toward harvest. To confinement producers, days on feed doesn't mean improved genetics, but rather less overhead and hours of care that has to be paid for by the corporation.

The modern confined hog may actually be a hog in little more than name only. Fat cover has been bred off of them to the point where they have to be kept in an artificial environment, thus they are far more prone to stress-related maladies. They have likewise been selectively bred for a degree of docility that makes them even poor automatons. In the end, they often go on to produce a pork product that, like themselves, is a pale and poor representation of what they once were.

The life of a hog inside a confinement unit can in no way be considered pleasant. Day in and day out, it is a never-changing scene; there is unrelenting crowding and other sources of stress, surfaces are hard and unyielding, even spilled feed can grind into skin and hoof pads, and there are all sorts of respiratory assaults inherent in this type of enclosed rearing system.

The Outdoor Option

Since taking the hog back outside is getting it back to its roots, it is not exactly a process of reinventing the wheel.

Despite efforts to reduce breeding stock choices to a handful of corporately held lines based on the Hampshire, Yorkshire, Duroc, and Landrace breeds, a fairly broad and deep swine gene pool still exists. Hogs that are well suited to the great out-of-doors can be found and should be propagated further. Much of the methodology for outdoor production was familiar to our grandfathers. One-sow farrowing houses, for example, exist in literally dozens of styles and constructions. Some now even incorporate space-age materials, a trend reflected in much of the equipment now available to range producers.

Feeding and watering equipment has gotten much better, and fencing options are perhaps more numerous and varied than they have ever been. In outdoor situations, energy and facility costs are generally greatly reduced, since the animal can indeed do much to care for itself.

Long ago, my granddad adopted a good farm tenet that bears repeating. He would state emphatically that he never made a dime with buildings and equipment, but rather with the animals passing through them. Confinement producers of late have seen their buildings depreciate to the point of being worth scant pennies on the dollar, and farms with existing waste lagoons are now steeply devalued, if they are marketable at all.

The hog out-of-doors benefits from a varied and changing environment. It can exercise and move more freely, thus improving muscle tone. And it benefits from a variety of outside stimuli and can more fully exercise natural behaviors and instincts. To borrow a phrase from a well-known dairy promotion, it produces the "pork from contented hogs."

A sow that has been freed up to do even a modicum of site selection and nest building will go into the farrowing process with much less stress. In a more open situation, she will milk better, eat and drink better, and should return to her estrus cycle quicker and in better condition. Granted, if left totally on their own, there are sows that will plop down in the nearest draw just ahead of a summer storm to farrow. The sooner you identify those animals and send them to town, the quicker that kind of behavior is bred out of a herd. For the range producer, natural selection of this sort is a very valuable management tool.

In nature, such careless and dead-headed animals and their offspring would soon be weeded out of the natural population. It is hard to measure to be sure, but lately it seems that confinement rearing has been shown to be little more than a crutch for some very suspect swine genetics.

The Big Question: Why?

Well, why this book? Why now? A lot of folks have simply written off swine production. They've chalked it up as one more loss to the "get big or get out" crowd.

Production agriculture can accept only a very few such concessions before it ceases to exist in any form we might reasonably recognize. Chickens were let go because farmers still had hogs, cattle, and row crops. Hogs slipped and some said, well we've still got beef cattle and row crops. Some, however, began to take note of the very few out there actually feeding cattle and processing beef and saw no comfort in that scenario. Let livestock production go and you have no option with your row crops other than to sell them for whatever the market offers at the time.

In the mid-'90s, butcher hog prices fell below 10¢ a pound for the first time in decades. It launched a bailout of swine producers of near-epic proportions. Hardest hit, however, were what would be termed the full-time hog producers operating with 100 to 300 sows. Just a few years earlier, these were the farming tigers, with the ideal family farm-sized — swine unit.

These folks had bought—and bought deeply—into the argument that to succeed you had to be a pork production specialist. They had that rug pulled from beneath them in a most sudden and ruthless fashion. And the explanation was that commodity pork specialists though they may have been, but they just weren't "special" enough.

At about 100 sows, a hog herd becomes an unhandy and rather unwieldy 1½ man operation. And the arguments of efficiency of scale were at last proven not to kick in until somewhere in the 3,000 to 5,000 sow range. Efficiency, evidently, could be measured only in boxcar-sized lots.

As one local wag so succinctly put it, "At 25 to 50 sows, a lot of folks were making some money. When they made the jump to 100 sows, it took all that extra money just to be what the slick farm magazines called a 'hog man.'" The added income from the additional 50 sows largely went to pay for the help and facilities needed to care for them.

It has often been pointed out that all that was needed to keep Premium Standard Farms out of the Missouri-Iowa region was for more farmers in that area to take up 50-sow herds and produce in an orderly form for that same market. The demand for good pork has never been a mandate for farm-crushing, environmentally degrading sow factories.

I find it interesting to note that one of the most successful alternatives to confinement swine production has also sprung up in this same area. The Neiman program boasts a substantial number of Missouri and Iowa farmers producing additive-free and humanely-reared pork for the lucrative West Coast trade. It has spawned a number of similar producer groups and alliances, and nearly all report that their greatest need is even more like-minded producers. Same have derided them as near-Luddites, but they're the ones who are setting the standard for premium pork in that region and throughout the country.

A number of Midwestern farmers are now direct-marketing range or "traditional" pork from their doorsteps and at farmers markets and netting $100 and more per head as a result. They have taken the classic whole-hog pork sausage product and used it as a vehicle for markedly increasing the income from swine ventures that ultimately are both more humane and better on a human scale.

Even this downshift in scale is enough to turn corporate heads. Awhile back I was on my way into a local farm supply store when owners of two of the largest remaining swine operations in our area stopped me to ask about activities at our own farmers market. The sausage there was readily selling for $2 to $3.40 per pound, depending upon whether it was sold as simple bulk sausage or specially seasoned links. Their very first question was how many hundreds of head they could expect to sell in the same way.

Ah greed, the great undoing of the American family farmer. I had to tell those men that this was a market almost exclusively for the 10-sow-or-less producer. Such sales can only be made one-to-one and one at a time. Most of us can, in a short time, list 50 to 100 folks who would potentially buy some of their pork needs from us. It is not a market for trailerload lots, but then that market has not been a true and dependably profitable one for quite some time. By the numbers, 50 head netting you $100 each are to be far preferred to producing 1,000 that net just $5 each. Simply put, scale can assure neither efficiency of operation nor surety of income.

Range-produced pork is now the porcine equivalent of caviar, and as such, it enjoys a premium—although somewhat modest—market. It is sought out by the informed consumer concerned with the issues of production. Buyers also tend to be those who have disposable incomes and can reward farmers who care most about what they produce.

Some Further Thoughts

The death of the independent hog producer is far from imminent. By some estimates, as much as 20% of all butcher hog production will forever remain the task of independent producers. At the moment, you may have to make an appointment a few days in advance to sell hogs at some interior markets, but a rebound is slowly occurring. As they have for the last several decades, the small producers simply pass under existing tracking radars and continue to produce through all of the ups and downs of this farming life, sometimes with very few animals and sometimes with a few more. They are that little statistical hiccup that has always bucked the trend, accounting for some degree of uncertainty in all of those USDA "Hogs and Pigs" reports.

The reasons for the assured survival of the independent producers are many and varied. Among them are: (1) there are a number of rewarding niche and small markets that are just too small to draw corporate competition, (2) where market outlets continue in place, small and established independents will endure, because of their low costs of production, (3) the corporations need the independent's golden image with consumers as a buffer of sorts against their own standards of production, and (4) the independents retain a superior degree of flexibility that enables them to more easily adopt all sorts of marketing and production options as they evolve and, in the end, are proven to be truly progressive.

I've seen 8¢ and 60¢ a pound butcher hog markets, and both can be survived. In times of extremely low prices, the traditional means of farm survival has either been to produce more or produce for less. Confinement-production technology has hit the wall. Little more can be cut away unless you propose to grow pork cells in petri dishes (and some of us wouldn't put that past them). And when you have 6,000 to 10,000 sows on some farms and CAFO (Concentrated Animal Feeding Operations) laws popping up everywhere, the numbers option has about been maxed out, too. The third road is, well, to carve out a whole new road.

Range production offers perhaps the greatest number of cost-cutting options now available for a variety of production choices. Furthermore, the simpler the total production equation is, the more enduring and profitable it will be.

Hog production today has taken a very different turn that few if any of us actually expected. Change comes quickly, and issues such as the environment and humane concerns weren't even conceivable nearly 40 years ago, when I began in hogs. It remains a business enterprise, but one with overtones only just now being realized.

We are poised before new and unique challenges: some swine breeds actually facing extinction, a purebred sector shaken loose from its leadership role and struggling for identity, a consuming public at once avid for the pork it remembers yet fearful of the product it is now being offered, an aging producer base, a rapidly disassembling support infrastructure, and numerous other causes of uncertainty.

The Chinese curse, "May you live in interesting times," comes quickly to mind. These are also times of opportunity, however. Contrarians of many stripes are emerging and finding openings and challenges aplenty.

Going back along the trail of time, we can find the seeds of early range production sown both richly and widely across the land. Early settlers were herding droves of hogs across New England up and down the Natchez Trace before they were rounding up cattle in Texas. With the growing importance of multiple uses for farm inputs and the stacking of compatible enterprises on even the smallest of farms, range production is a concept whose time has come again. And it has come again for the better.

Range farming is a type of production that will require an artful blend of genetics, facilities, and producer skills to succeed. It won't be to everyone's liking, and it carries with it now a fair bit of controversy. It won't pit neighbor against neighbor, but it does concede a fair number of points to some of agribusiness's more vocal critics.

For some, it will look like nothing more than a step backward to knee-boots and 5 gallon buckets and choring out in the elements. Yet with it is the promise that at the end of day, there will be black at the bottom of the ledger page, our neighbors will again be just that, not competitors, and there will be that good feeling of a job well done and done for all the right reasons.

The Lowdown on Today's Pork Producers

Today there is an array of different types of producers canvassing the swine production field. There are purebred producers, swine finishers, contract producers of weanlings and feeders, a modest number of farrow-to-finish producers in confinement operations, and a growing number of what might be termed "generalists" working independently and with modest numbers of animals.

The producer most under fire, perhaps surprisingly, is the farrow-to-finish producer working with confinement units. A few short years ago, these were the lions of the industry, but they were also among the first to fall victim to the new numbers game being played in swine production. Breeding herds, farrowing units, nurseries, and starting/growing units in the numbers required now are no longer simple or easy to fit on a single farm.

The minimum of 3,000 sows needed to be considered a player in the porcine industry doesn't call for a hog farm, but for a hog factory. They require the work not of a farming family, but rather a crew of highly trained swine technicians. And such people are more suited (and interested) in a 40 hour workweek than the traditional 24/7 schedule mandated at certain times of the year with livestock farms.

Most of these producers who remain are operating at the 150 to 250 sow level (one to two person ventures) that now have a hold on the land that's tenuous at best. They are generally fairly heavily debt-leveraged and locked into place at a hard, fixed level of production, and at a level of technology that daily becomes more dated and worn down. Their marketing outlets and options have also grown far fewer in number in recent days.

Such producers in our area now can only sell to just one general stockyard, and one packing plant pays strictly on lean yield—and both are based in western Illinois. Cull sows go to a single plant in the far northwestern part of the state. Some local volume producers have made links to producer groups across the Midwest, producing specialized products, such as antibiotic-free animals or pork based on the prized Berkshire or Tamworth breeds. These options have required long shipping distances, investments in often difficult-to-locate seedstock, commitments to set numbers based on very narrowly focused markets, and frequently, producer investment in packing facilities.

Some others in this category have opted to go the contract route, usually taking on a very specialized role. They may opt to produce a company-selected pig weaned at 10 days of age or so, or early-weaned pigs being backgrounded and developed in hot nurseries, or finishing the pigs in droves of a thousand or more delivered to their buildings. They do this at a flat fee per head, and with very few and modest premiums for any exceptional animals.

Their contracts are seldom more than a year in duration and require buildings that may not be paid for in even a decade of constant, heavy use. The contracts generally offer the corporations all sorts of early outs and may even punish the farmer/producers for any sort of independent action. Efforts to organize contract producers have met with limited success, and there are many instances corporations canceling their contracts with producers, sometimes giving little advance notice.

Many producers see contracts for production as just one more cudgel with which the family farmer can be beaten. They are supposed to bring a more businesslike structure and orderly supply to livestock production. Well, they do help the corporations give individual farmers the "business," but all they really do is continue to encumber farmers with risk and debt obligations while freeing up the corporations to go their merry way anytime they choose. A contract may be an "agreement" between two parties, but its real result is to generate a form of captive supply for the packing industry.

The purebred sector is now very greatly reduced in size, and those farmer/breeders remaining in place may constitute the oldest group of swine producers around. Farms and herds three and four generations—producer generations—folded coming out of the mid-'90s, and those that remain are functioning at far different levels and with far different production mixes.

That mix now includes bred and "open" (mature but unmated) gilts, fancy butchers, show pigs, and a handful of young breeding boars each year. One noted Hampshire breeder in our area had to draw a line in the sand. He has spent a lifetime building his herd and refuses to sell young boars for less than $500. His yearly sales of such animals are now a fraction of what they were a decade ago, but he reports recent demand as steady and growing. His numbers are now such that they would constitute a quite profitable niche market on nearly any family farm.

At the swine shows at state fairs now, you will see a lot of 4-H and FFA youth, a few old familiar faces, and surprisingly, a varied and growing group of newcomers—folks with small herds of multiple breeds, preservation breeders, those being drawn to the show-pig trade, and some commercial producers now maintaining a purebred herd to produce at least a portion of the replacements for their commercial breeding herd. A neighbor of mine now has a small herd of Tamworth hogs to supply boars for a special Tamworth-cross program in which his commercial herd is enrolled. An unexpected benefit is that he is back out showing a few purebred hogs at the fairs, just as he did years ago when still in the FFA.

It is hard to draw a parallel of any sort to what is going on with purebred hog production right now, but the purebreds do seem to be at or near a position that has been held by purebred sheep for roughly the last quarter of a century or so. They are in more than competent hands, but there is a slight element of uncertainty about them. That they can be put to the greatest possible number of market uses is all to their good, and not only will their value and use continue, but will grow.

For the moment, they are in what might be termed a state of transition. In the near term, there will be a changing of the guard as the older producers step down. The new generation of producers will then have much to say about how their animals are to be presented and used. Perhaps even competing breed groups may arise, as is now being seen in the beef cattle sector.

The generalists are coming upon the scene in greater numbers now. They are direct marketing hogs and pork to a great many sectors and are finding their way largely from inside the existing pork production infrastructure. From a small herd, they may be selling everything from breeding stock to fancy cured hams, and they are always looking for ever more marketing options. They are also fitting their hogs in and around other ventures on their farms.

Range rearing is the production option that holds the brightest promise for the independent farmer. Into a mix already rich with grass-fed beef, range broilers, grass lamb, free-range turkey, and pastured eggs comes the range or outdoor plain ol' "dirt hog"—nothing fancy, nothing overly hyped, just the gathering of available swine genetics to create a new/old hog and pork.

The range producer can pursue any or all of the market options that appeals to him or her. It will most likely be a good mix of them all.

I can envision a time when America's farmers—her family farmers—will come to embrace a school of farming of a very artisanal, or craftsman-like nature. It will be creative, humane, environmentally sound, and have both strong financial and moral integrity. The outdoor producer, the artisan of meadow and paddock, will emerge with the image and the regard now held by the best of the vintners.

It will be no simple undertaking and may still be the better part of a lifetime away, but its roots will be in a more natural and fulfilling type of production. From it will come meat animals every bit as distinctive as fine wines—and that are valued as highly.

The artisanal swine producer will work with modest numbers, always emphasizing quality over quantity. The volume producer in any of the categories above has had naught but the wholesaler's example to follow. Their answer to every problem has been to crank out ever more product, and in that particular structure it will always be the only possible response until all is washed away in a sea of unsalable product.

The artisanal producer raises an animal to meet the needs of some very specific markets. Once upon a time, hogs filtered out from a few rather specific areas to meet the needs of both regional and national markets. Older cull breeding boars from the Midwest, for example, often found their way to specialty sausage producers in the East and Northeast. A Missouri boar arriving on the loading dock of a New Jersey packer may have actually traveled farther from home than his previous owner ever had. And these were markets served by animals often bought just one and two at a time in the field and at interior markets.

Now niche marketing has grown so much larger and includes so many more options. By their numbers, they will never require huge numbers from producers, but when they are assured of dependable sources of supply of the kinds of the animals they want, they can become quite enduring markets. Each emerging range producer must set a target number that will do two very important things: (1) serve markets that will reward production with fair and regular returns, and (2) be at a level that the producers feel to be a comfortable fit. Those who are driven by circumstance into production never truly produce well.

We built our purebred herds to levels of no more than eight to 10 sows and found that over time, the right five to seven sows could produce the 35 to 40 purebred boars that we could readily market each year. We addressed our numbers needs by selecting for sows that farrowed and raised good litters and that were genetically strong and deep. Numbers should thus be attuned to the reality of your market's needs, rather than some nebulous quest for ever upward-spiraling numbers. Growth simply for growth's sake, after all, is the philosophy of the cancer cell.

In my years upon the land, the necessary number of producers often seemed to be wafted down from on high, trumpeted by county agents and slick-paper magazines. They were spun out rapidly and ever upward in the last four decades of the 20th century. To fit the times early on, you needed at least 25 sows, then 50, then 125 (I never figured out how they came up with that number), then 200, 300, 500, 600 (which was a bit of a plateau), and now farms are said to need 1,000 to 3,000 sows to be deemed efficient. This is America, and I would hate to see caps or restraints applied to anyone's ambitions, but with much beyond 50 sows on a family farm, you have to start adding sows just to pay for the care and needs of the previously added sows. I think a lot of volume producers would be surprised to dig into their records and determine just how few sows are paying directly back to them.

What we see now are a growing number of six- to 12-sow herds and a good many folks with only two or three sows tucked away in a wooded lot or small pasture. These folks don't show up in studies and surveys, since they are largely cash-and-carry buyers who tend to shun producer groups and government programs alike. A lot of producers who were said to have gone away in recent years actually just scaled back rather dramatically and opted out of the pork producers scene.

If it is the case, as I and many others believe, that independent pork production is on the comeback trail, it will have a form that is very different from the one in place only just a few years ago. Simple commodity pork will continue as a presence, but pork producers keying on consumer concerns will grow in numbers and importance. The role they will come to play will be limited in importance only by the input they are willing to make into it.

The individual may be supplying one butcher hog a week to a local restaurant, growing out 50 or 100 butchers a year for a livestock-based CSA, or building and promoting a purebred herd of Berkshire or Hereford hogs. Successful producers will begin creating with their hearts and minds, not by simply generating mere numbers.

POLAND CHINA

CHAPTER 2

HOUSING & FENCING

It has been said often and rightly that I was born to be a livestock farmer. I'm most at home in sale barns and along small town streets, where the talk is always of crops and weather and flocks and herds. And yet I always try hard to stave off "barn blindness" when it comes to country and rural ways and people.

Many years ago, I played host to a team of Eastern writers for the Rodale folks as they did a series of interviews with Missouri hog producers. The photographer was especially enjoying the trip and commented frequently that although the hogs required a greater physical effort to get just the right shot, they were still easier to work with than the fashion models he had been photographing a few weeks earlier.

The real high point of that trip, though, came a few hours later in the day. We were sitting around a picnic table a short distance from a turn-of-the-century clapboard house that was home to a three-generation farming family. The third generation, an FFA member justifiably proud of the Chester White herd roaming the nearby pastures and wooded lots, generously shared his grandmother's home-baked cookies and his mother's iced sweet tea. The hogs were at once both a natural part of and a funding engine for the good life enjoyed by this very special American farm family.

It was a true Norman Rockwell moment and yet one easily clouded with excessive sentimentality. For within it was one of the greatest of all pitfalls for farming folk. If at the end of the day, income does not exceed expense, then all of our good work has been for naught. All of us out here on the land value the way of life, but that way of life must have a sound economic base if it's to endure. The family farm must also be the family business, operated on sound business principles.

Since the age of 10, and except for a short time away at school, I have never been out of the line of sight of a growing crop of corn, or corn in one form or another in storage as a feedstuff. Its warm, mealy smell is second nature to me, and the clang of feeder lids and the bump and clatter of penned hogs have been a gentle background for much of my life. In the old storage box behind our heating stove, you would as often find little pigs or other barnyard infants warming as you would gloves and caps drying. We did our best to make their lives safe and comfortable, but we did so keeping in mind their need to be real hogs for a real world.

To succeed with hogs, you have to like working with them. But beyond that, you have to see them fully, the good and the bad, and to be able to operate with a clear plan for production and marketing. They are endearing creatures, but in no way can you justify the cost of keeping them simply as companion animals for a farm lifestyle predicated on driving big tractors and regularly trading 4 x 4 pickups. To be a success, there must be a clear "why" to any farm venture, and that means profit and suitability mean far more than numbers, image, or having hogs to grow corn to drive big tractors.

Until well into my 20s, I was known largely as "the kid with the good red hogs." It was not a bad way to be thought of, but it also meant that by the age of 16 I had a checkbook and feed bills, and both had to be reconciled.

The farmer considering a swine venture now must ponder the variables of the market, the suitability of the enterprise to the home farm, the necessary producer skills and interest, and the risk. This "corn-er" has long been an advocate of enterprise diversification to assure farm stability and farm business success. Indeed, it is virtually impossible now to find a single farm venture with the earning potential to meet all of a modern family's needs.

Furthermore, the costs to establish a single venture with even the remotest possibility of generating a viable income for a family are all but totally prohibitive. A 10-venture enterprise mix netting just $2000 per venture would generate a yearly income of a rather respectable $20,000. And a little tweaking and tinkering to increase the net just

$500 per venture will increase overall income by a most respectable 25%. Ventures of this scope are simple to establish, easily financed, shouldn't disrupt an existing enterprise mix, and if necessary, can be folded without taking out the whole farm.

Making the Start

I am a firm believer in doing all your early farming on paper. Gilts and boars should never, ever be impulse buys.

A first assessment of the viability of a range or outdoor swine venture should ask the following: (1) Is labor available? (2) Will it compete directly with other new or existing ventures for available resources? (3) What are the available marketing outlets and options? (4) What kind of support infrastructure is available (*e.g.*, vets, feed suppliers, seedstock sources, etc.)? (5) Do you like working with hogs? and (6) Is the family in accord on the venture?

For several decades now, swine production has labored under the concept that you have to be a very focused swine production specialist to succeed. You must either be a producer of seedstock, a feeder pig specialist, or a feeder/finisher of hogs. This sort of thinking has never really made sense to me, nor is it in keeping with most of the more successful industrial models. Walk through a large Wal-Mart store now and you will be offered everything from oil changes to summer dresses to pork chops.

Historically, there has been a certain degree of regionalism to U.S. livestock production. It has been based on cropping patterns, climate and rainfall, access to shipping and population centers, and some fairly unique variables such as the ethnicities of nearby populations of consumers. These factors still remain as relative constants, but some changes are happening. At this very moment, an effort is under way to create an animal agriculture model in northern Iowa. The grain source is there, and a Hasidic Jewish community has arrived to better funnel grain-fed beef East, and dairy farmers from Holland are being recruited to launch a new dairying zone.

Hogs certainly have a strong Midwestern and Southern identity, but as *Lonesome Dove* illustrated, old sooeys can soon settle in and call just about anyplace home. In fact, Midwestern producers are given undue credit for some of their supposed role in pork development and production. The still very important Duroc swine breed was originally called the Duroc Jersey, because of its development out east in the Garden State, New Jersey.

A few years ago, a Midwestern barrow-show jockey of my acquaintance regularly traveled to the Cow Palace show in California and was most handsomely paid for the pickup load of hogs he chauffeured west. Of course, they really weren't any better simply because they had drawn first breath in the Corn Belt, but the perception out there was that they were.

Gene pools may be broader in some areas and grain sources more abundant, but hogs will fit somewhere in all 50 states. In the course of a Missouri year, temperatures can range from 102 F to 20 degrees below. Such extremes will take a slight toll on performance, but they make the adjustment, and savvy producers keep them performing profitably. Even in the island paradise of Hawaii, there are farmers producing roasting pigs for that classic meal, the Hawaiian luau.

A range swine venture will have busy times and other somewhat slower periods. The outdoor producer is most apt to be following a seasonal pattern with the high workload times of farrowing coming in early spring and early fall. March, April, and May farrowing females can be bred to farrow again in September, October, and November. A seasonal pattern of production used even earlier was to farrow gilts in late spring or early summer, wean their pigs at 6 to 8 weeks of age, feed out the gilts to a good flesh, sell them, and replace them the following year with their daughters. It would pare some overhead costs and have pigs being born and growing only in the most favorable weather. The conditioned gilts also held substantial value each as replacements for other producers or as slaughter stock.

Outdoor produced pigs will generally nurse for 5 to 8 weeks, unlike the 10 to 21 day range now proscribed by many raisers in controlled environment facilities. Lactation is a very critical time in the production calendar and can set the course for the animal's entire lifespan. Most never fully overcome a poor start. Early weaning—21 days or less—is a high-tech sort of thing at best.

At approximately 21 days following farrowing, the sow's body undergoes many of the chemical changes that would normally accompany her estrus cycle, and this can create a window of stress and disruption. Many believe it's the source of what has come to be known as the 3 week or 21 day scours in nursing pigs. (Symptoms include white, pasty scours rather than the watery variety. The pigs may be able to shake it off in a day or two, even if left untreated; however, this could affect productivity.) For the best performance, natural cycles must be respected, since rushing the sow into reproduction can take a toll on her. Sows in confinement now seldom last through even four parturitions.

Fortunately for farmers wanting to fit swine production into their enterprise mix, sows can and will breed to farrow in any season of the year. Even with a seasonal approach, sows can generally be bred to farrow ahead of or behind most rush periods of fieldwork or around other livestock needs. The labor investment per year for a sow and her two litters is on the order of 20 hours or less.

Be careful, however, not to buy into that argument that if you are going to take care of a couple hogs, you might as well make it a couple dozen. Throwing protein cubes over the fence for two sows or 10 takes about the same time, but times like that are few. You tend most litters one pig at a time, and the time demands for 10 litters are 400% greater than for two.

Hogs reared outside were long the foot in the door for young farmers. They respond well to an investment of sweat equity. The more care and focus they are given, the larger their litters should be and the sooner they will breed back. The more timely a management practice is given, the greater the response from or return to it will be. From one Duroc gilt bought early in my junior year in high school, we grew to six females by the time I graduated. Working in and around extracurricular activities, and even with time away at college, we were up to 25 sows 2 years after that.

Confinement rearing can make speedier work of some tasks like watering and filling feeders. However, the all-important tasks of baby pig care and managing the breeding herd still require a hands-on approach and an investment in manpower. On a per-sow basis, there will be a modest increase in labor investment on range over confinement, but with moderate numbers it will seldom be significantly greater. Furthermore, this is often a labor investment that can be made at off hours.

Range production of hogs will fit onto most of those small and oddly shaped parcels of land that can be found on nearly every farm. They should not be turned directly to ponds and streams, nor into woodlots producing timber. Granted, old-timers would let hogs go on ponds with defective and leaky bottoms, relying upon their wallowing activities to smooth out and reseal the bottoms.

Hogs can foul water, making it unsuitable for drinking, and create a degree of turbidity that can be harmful to fish and other aquatic life. Rooting activity and hoof traffic can keep banks muddy and unstable and can also do damage to the pond's berm. Still, they can do a brush-clearing job every bit on par with brush goats, and they are a good fit for rolling, lightly wooded plots. Hogs and the gravelly hills and ridges of eastern and southern Missouri are a truly natural pairing. It was this fact that played such a key role in the survival of the grand old breed, the Mulefoot.

One of possibly the last three Mulefoot breeders left was and continues to be based in eastern Pike County, Missouri. His hogs were kept in wooded areas near the Mississippi, and even on river islands. The breed hung on quite literally because those animals were surviving in an environment that would have been familiar to their feral ancestors.

Hog housing can be fairly easily reworked for use with a number of other livestock species, but is not easily shared with other animals. Feed storage, transport, and handling equipment, fencing supplies, and more are all usable with any number of other livestock varieties. The additive-free beef producer or range poultry producer will see outdoor pork production as a natural fit and a new product option for their overall scheme of production.

Small-scale hog production systems fit well because they can be set up so that they don't compete for producer time, farm space, and other farm resources. We have raised hogs with every other major livestock species in our farm mix and found the fit to be a good one as long as we carry out the needed planning and forethought. Having broiler chicks arrive on the same day pigs are weaned, for example, can be corrected through simple scheduling. Having 200 pound shoats and chicken tractors made from PVC in the same pasture is a formula for disaster. There are no major livestock or poultry species that cannot coexist on the family farm, but like any other type of matter, they cannot occupy the same space at the same time.

A perception that has been dogging the small-farm movement for generations is what I have come to call the "big red barn fallacy." It is the belief that at the hub of every farm, there has to be this giant building that is home and haven for everything from doves and banties to draft horses to all manner of barn cats and fuzzy little wild critters. Talk about a recipe for disaster.

What doesn't get eaten or trampled to death will languish in an environment that isn't designed adequately for their individual needs. The costs for huge buildings, not to mention their space-consuming nature, have put off far too many families from pursuing livestock ventures. I have seen $50 roosters drink contentedly from old tin cans, and old bed springs used for hog panels.

I know of producers who are lambing ewes in January, farrowing sows in April, and planting corn in May, and they still have order in their lives and manage to stay civil with their spouses. On even the smallest of acreages, a number of ventures can be stacked into place as long as they are not directly competing. The flexibility of a range-swine venture is certainly conducive to many enterprise mixes. The marketing options, from feeders to butchers to seedstock, afford timing options that can both even cash flow and assure good labor utilization across the full calendar year.

The End Product

The key question to ask of any farming venture or pursuit is, what are you going to do with its output? Does a market exist that will recognize your good efforts and reward them with a fair price? I don't believe that farmers must live or die with the old "sell it or smell it" marketing mentality. Nor is it good business to wait until you have something to sell and then ask, what will you give me for this?

Markets give nothing. A solid rule of thumb for any farm venture is to know where you're going to sell its output before you buy into it in any way. Pork, the commodity, has fallen to the vertical integrators. They will buy a certain amount of product from independent producers to cultivate the image of pork production as still being in the realm of consumer-favored family farmers. The ups and downs of the marketplace and the infrequent, modest returns to the on-farm producer are reflective of what the traditional marketplace has become. In a vertically integrated system, the product will pass through any number of handlers before the true profit is taken at the last point of corporate control, the retail sales counter. And that, my friend, is a mighty long way from the farmyard gate.

Fortunately, a number of marketing options—some very new, some very old—exist in the wake of the vertical integration movement. Perhaps first among them is pork produced for the tables of an emerging and growing number of consumers with strong health and environmental concerns. They are a segment of the consuming public that has also expressed a willingness to pay something of a premium for the type of production it values.

Other niche markets now include show pigs, feeder stock for sale to other modest-sized farmers, roasting hogs and other specialty meats, seedstock, organic production, and more. Much hog production is now sewed up by contracts, which leaves a great many people wanting pigs and pork but with minimal outlets to turn to for their needs. Only relatively recently, however, has there emerged anything like organized efforts to link these consumers with independent hog raisers.

Now and for quite some time to come, producers will have to expend a substantial amount of time and effort on the marketing process. It may even have to rival the time expended on actual production. In later chapters, we go into greater detail about the various marketing processes now open to the independent swine farmer, but for the moment, bear the following in mind:

1. Markets will be what *you* make of them—they are often producer-built, one pig at a time—and direct marketing will be your key to profit and survivability.

2. You must emphasize quality over quantity.

3. To survive and profit, you will have to take on a cutting-edge focus, marketing to buyers who have special demands and are willing to pay extra to have you meet them. The sweat of your labor, the feed you feed, and the kind of care you give will amount to far more than numbers and old-school performance data. Pork as a mere commodity has left the family farmyard just like Elvis left the building, and in its place is hog production for very special and exacting markets.

It used to be said that Swift Packing Co. had a way to sell everything but the hog's squeal. The range producer will be selling the hog and more, particularly how it was produced, by addressing the concerns of a now well-informed consuming public. Your market is just as apt to be shaped by the latest report on *60 Minutes* as it is the most recent USDA "Hogs-on-Feed Report."

The Oh So Very Important Little Things

The current state of affairs would once have been unimaginable, but in many rural areas now, very little remains in the way of a support infrastructure for hog raisers. The number of large-animal veterinarians has gone into marked decline, feed suppliers are fewer in number and more limited in their choice of feedstuffs, and to line up good seedstock now can mean a trek across many miles or even many states.

I spent many years in the feed industry and can recall feedstore warehouses filled with literally dozens of different types of swine feed for a wide array of rearing and feeding programs. Most feeding programs are currently drawn up for hogs in some sort of confinement system. As such, more starters are for early-weaned pigs, growers are for hogs with the controlled environment crutch, and sow feeds have been likened (justly) to jet fuel.

Wherever possible, I like to encourage what might be termed a loose networking of area swine producers. When farmers can again view each other not as competitors but as neighbors, good things can and will begin happening. As a group, they tend to hold suppliers in place and to draw more buyers to the area. They can often share in the purchase of some inputs, and perhaps most important, they can provide each other valuable shared experiences and support.

A community of like-minded producers can do so very much for each other. For example, although our little town is small, area farmers provide enough business to support two local elevators. As a result, they compete with each other a bit, and we in turn reap the benefits. I have seen the elevator west of town sell shelled corn for as much as 5¢ a bushel less than the elevator to the east of town. Such competition assures everything from freshness of feedstuffs to being offered the latest in health supplies.

While you wouldn't want to start a canary ranch high on the side of one of the Rockies, you also shouldn't expect a swine operation to fare well where the local vet sells five kinds of cat shampoo and hasn't seen a hog since his college days. A community of like-minded producers is needed to encourage the formulation of a like-minded support infrastructure—something that's always needed for producer success.

Last, but maybe most important, is that your heart has to be totally in it, and the support of the whole family is a must. The range producer these days is to a certain degree embracing the role of the maverick—this ain't the way your daddy or the land-grant colleges raised hogs. Your granddaddy, maybe.

You will need to form the mindset needed to take you along a uniquely independent path. Small farmers, niche marketers, those using range production, and just about anybody else outside the high-tech loop takes a lot of heat about being backward thinkers and anti-progress. You have to consider "unconventional" approaches. Several years ago, an acquaintance of mine was advised by a buying-station manager to upgrade his production. Instead of buying a couple better boars, which would have been both the appropriate and quickest fix, he went out and dropped five figures in a state-of-the-art farrowing house. Of course this didn't make the hogs coming from it one whit better, but it was considered "the progressive upgrade." Sometimes the epic changes proposed by agribusiness fall far short of the real mark of making you a better and more successful swine producer.

When accused of this sort of "regressive" thinking, I take great comfort from an old adage from the French that says, "When you come to an abyss, the only safe step is backwards." I can't think of a better image to describe the state of affairs in swine production of late than a great abyss.

Focusing on Free-Range Production

As stated several times above, hog production will fit onto a great many family farms and can make meaningful contributions there. Outdoor production has the added punch of being a natural fit to many of the currently hot livestock ventures: range broilers and mobile housed egg layers, grass-fed beef, organic production, additive-free meat, and community supported agriculture. Outdoor pork is the logical choice for CSA producers who are already marketing produce and poultry products directly. It can go a long ways to helping fill in the off-season gaps that occur with market gardens and the like.

We should take a minute to be straightforward here: hogs are not the best utilizers of pasturage. They are not ruminants; they have a straight gut, unlike the ruminant. A total legume pasture mix won't harm them as such, but accepted belief is that a growing/finishing hog will meet about 10% of its total nutritional needs from pasturage. A sow will probably be able to meet 20% of her needs on good pasture in season.

The old standard for stocking was 10 finishing hogs per acre, and at this rate they were meant to feed heavily on the pasture crops. This figure can be bumped if the hogs are being offered a good growing/finishing ration free-choice. A 15% crude protein feed will easily double or increase this number by 150%, but be careful that stocking rates don't begin to damage pasture surfaces.

As many as four sows and litters will find room on a quarter-acre plot if the sows are also on a good gestation ration that's offered free-choice. Old hands tended to favor six to eight sows and their litters per acre of lush pasture, but again with a good sow ration fully fed. The first and best choice of hogs to put on pasture is sows. Gestating sows are a natural fit for the range year round, and also for gleaning corn fields following harvest.

Thin sows, when they can be found, are eagerly sought out by growers needing to glean grain fields. In fairly short order, they can make 75 to 125 pound gains if they're thin but otherwise healthy. Several years ago, we sent a group of thin sows to a local buying station, and we prepared to swallow hard and take a price beating. The operator there rerouted them to a nearby sale barn, where they brought nearly double the slaughter price for killing sows in order to go back to the farm to glean behind a combine.

When sent to glean harvested grain fields, shoats greater than 70 pounds or gestating sows can meet the bulk of their nutritional needs. The more grain that remains in the field, the better the hogs will fare, and in the early going, they may need little more than water and a modest bit of mineral and protein supplement. With hogs out gleaning, the producer must walk the fields often to determine how much actual feed remains available to the hogs and the level to which supplementation should be increased.

This gleaning of fields is a long-standing farming practice, one that no doubt launched the feeder pig industry. In very wet years, "hogging down," as it is commonly known, was sometimes the only way to harvest an entire field of corn. Indeed, before there were feeder pigs, there were "cornfield hogs." These were generally crossbred shoats that were 60 pounds or better and showed a lot of Black and/or Red breeding. Those sandy-colored shoats of old were a very valuable commodity at harvest and shortly after.

Hogging down still works and produces a very desirable type of meat animal for many consumers. An acre of corn will feed a substantial number of hogs if they are allowed to harvest it themselves. To grow a hog to a good butcher weight, you can still rely on the old rule of thumb that holds that 10 to 12 bushels of shelled corn are needed for the task.

A friend of mine, Ron Macher of Clark, Missouri, will turn out 10 head of growing hogs on one acre of his own open-pollinated corn. When possible, he will use feeders from one of the heritage breeds like the Tamworth or Hereford. He turns these 10 head into one and two-pound sticks of whole-hog sausage that he sells from his doorstep for $2.50 per pound. The price is more than justified by what he is packing into this classic staple of the American breakfast table. He is not simply selling pork sausage, but rather a rich blend of history, nostalgia, and breed preservation in a food item that is both "natural" and socially correct and that gives the buyer a stake in preserving things they value, including the family farm. It is locally grown, locally processed, direct-marketed (to attest to its freshness), and it tastes good. Who wouldn't want to fill the home freezer with all of that?

Outdoor pork production doesn't have to be ruled out on even the smallest of farms. Our bit under three acres has been a comfortable home to as many as eight Duroc sows plus the feeder pigs and up to 40 boars a year we produced from them. We don't have the freedom to rotate through a number of large pastures and lots, but over the years we've picked up a few tricks that enable us to keep them with the Sun warming their backs and fresh air in their lungs.

We take the farmyard concept several steps forward and endeavor to create a drylot that is truly hog friendly. At a minimum in a drylot, you need to provide a growing hog with at least 100 square feet of lot space, not including room for the sleeping shed. For sows, provide at least two to four times that amount; as with buying jelly beans, it is always best to err on the heavy side. Space requirements in drylots will vary greatly across the country, depending upon local rainfall patterns and muddy seasons.

Mud problems can occur with this system, and all feeding and watering equipment should be placed on large platforms. In an extremely muddy year, we have used creek gravel to fill especially wet areas. We also try to maintain a sufficient number of lots so that we can rotate out of any problem lots in very wet weather. Much can also be achieved with good lot design and layout.

Our lots are set up on lightly shaded, gently sloping tracts. The sow lots set just below the farrowing huts, which sit atop a gentle slope and alongside our county road. At the foot of each lot, we maintain a 20 to 40 foot wide strip of permanent sod. Here in Missouri, those strips are heavy on fescue, and its dense sod not only slows runoff, but filters it as well. The water emerging below the strip is slight and very clear.

For all but those fabled 100 year rains, the lots receive a scrubbing of sorts with each rain that has something of a renewing effect on those surfaces. The sod strips effectively filter any runoff and largely hold the lot in place. This methodology has enabled many swine producers to prosper in the cherished "river hills" that rise up along the Mississippi in the nation's heartland.

Running the Numbers

Budgeting is that piddly little chore that always drags dreams back to Earth. Putting dollar figures to your dreams is never easy, but you should always do your farming on paper first.

Figure that a 40 pound shoat will consume 10 to 11 bushels of corn and 125 to 150 pounds of protein supplement on its way to reaching a good market weight. A sow will consume 4 to 5 pounds of a fairly nutrient-dense ration daily during breeding and gestation. During the last third of gestation, that figure might go up by 1 to 1½ pounds, since this is when fetal pig growth occurs rapidly. During early lactation, this will gradually build up (over a period of 10 to 14 days following farrowing) to a full daily feed level for the sow; this is roughly 3% of her bodyweight. A breeding male will be held on what is basically a maintenance ration for much of the year, depending upon his body condition, and will consume something on the order of a ton of feed per year.

Feed will be the largest constant budget item in most range operations. There will be overhead items like interest rates and taxes that will also be constants, but feed will be the big-ticket item for most producers. And please, please don't fall into the trap of believing that because you grew the corn, your only loss in using it is the cost to produce it. If you can sell the corn you grow for $1.70 or $2.70 a bushel, that is what it costs you to feed each and every bushel of your corn. Buying corn or other feed grains rather than growing them has been an option for a great many swine producers over the years. This is in part because at one time, many producers set up their enterprises around the Corn Belt to funnel feeder stock into it, and thus they created a pattern of tradition. Others recognize that their skills and resources are simply better suited to livestock than to crops.

Veterinary costs can be pared through comparative shopping, and they can also be expected to shrink over time, as the producer gains experience in his or her diagnostic and treatment skills. My grandfather, Kelly Brewer, placed a high value on learning, and he considered the costs of acquiring or broadening one's skills justifiable at all levels of production. What's more, he was a farmer from the old school, and he found it difficult to perform any of the tasks that would have caused even mild discomfort or would have drawn blood from the animals, which he dearly loved.

When we first returned to the farm, there was a generation of men still in place who functioned as lay practitioners of certain types of animal care. The gentleman who did that in our neighborhood was Mr. Jack Motley, and he was sort of a cowboy who never quite made it to the plains. He could do castration work, dehorning, tail docking, and the like. The tools of his trade were largely a well-worn Boker pocketknife and a Prince Albert tin full of pine tar. He would castrate only when the moon signs were in the feet and legs, to be sure that the surgical wounds would drain properly. It might be a month after calling him before the moon and his mood suited him, and he would roll you out at the crack of dawn for the operation, but they never came too big or too rough for ol' Jack Motley.

I was still in my early teens when Dad decided that I needed to learn to do as much of my own vet work as possible. He called our local vet for a field appointment and set the two of us to working up our spring crop of pigs. Under the watchful eye of the vet, I learned to do castration work, give injections, and handle other small tasks.

Later, my father had me sit down with our lawyer and learn income tax preparation, what was and wasn't deductible in a farming operation. It taught me that even those fixed overhead costs can be managed.

Learning is always a good investment. In fact, it's what you learn after you think you know it all that will serve you best in a farming life.

I am in no way a tax expert, and I realize that no two farms are even remotely the same, but this much is very clear: range producers must have a bead on a good a business plan from the get-go. For example, the need for hog houses is a given, but hog house on runners won't appear on local property tax rolls as fixed, taxable structures. In fact, they will qualify for the rapid-depreciation provisions now in place in the federal tax laws.

The Thought Process

Early on I noted the risks of becoming too specialized in any type of livestock production. This was a hard lesson to learn for a person like me, who took a lot of pride (and a fair bit of my personal identity) from being known as "the fella with the good Red hogs." Still, just having hogs isn't enough; the good swine producer is focused and is always striving to develop ever further his or her expertise.

What, then, makes a person a truly good "hog man" or "hog woman?" Well, it certainly isn't being able to produce them by the tens of thousands or drive the biggest 4 x 4 from General Motors.

The successful hog person now, to use a good old term, is one savvy person. He or she has formed an honest appreciation of the animal, has acquired the necessary skills to raise it, and has resolved to make a commitment to quality and to be in the business for the long haul. Which is important, because an investment here is the same as it is for any other market. To have hogs to sell when they are high, you are also going to have hogs to feed and tend when they are selling poorly. The good swine producer is thus ever and always a student of the business and the animal.

They study their markets long term and short, they shop comparatively for all inputs, they track trends and know swine genetics, stay current on the literature in the field, and have learned to honestly assess their own operations and efforts. I can remember and even name a handful of boars that were popular when I was still in the FFA, but there are folks who have 50-plus years of bloodlines and lore in their heads. If you can remember boars named Boxcar, Liberty Bell, and Bulldozer, you are like me and have begun the countdown to retirement.

There are producers spending two hours a day and more keeping current in their reading and studying their farm records. The latter constitute "the book" that will keep you ahead of the game. This is the kind of time that is always very well spent and can literally return thousands of dollars per hour invested. Pick the wrong boar and it may be years getting back to where you were. And the guides to such important decisions are all part of the documentation of your venture's performance and your ability to gauge the current state of the industry. It's been said that a good hog raiser can do more with baling wire and a 99¢ pair of slip-joint pliers than Donald Trump could with a million dollars.

Good hog-raisers are the masters of the home farm, one that is like no other anywhere on Earth. They know what they need in their next boar, and generally the one after that. They know where to save a few pennies per bushel on corn and have a pretty good idea what the market will be doing 90 and 180 days down the road. Pork production isn't exactly like the game of chess, but you sure don't want to play checkers for money with a good hog raiser.

The Setup

You'll want to begin your range swine operation with a stack of graph paper, a pencil, and a good eraser. Lay out your farmyard, lots, and pastures as they currently exist. Then begin visualizing and planning how to have the animals move through this arrangement to achieve optimum performance.

Proper animal flow through the facilities assures cash flow back into the operation. And that flow must be calculated throughout the entire year. It is better to realize on paper that you'll have piggy sows and bred ewes vying for barn space in February than to wait till you've taken the leap and find yourself and the animals standing flank-deep in February snows.

A great many range operations work on the wagon-wheel design, in which all the pastures and lots are centered on the farmyard or some other central hub. At this hub, inclement-weather farrowing can take place, animals can be sorted and loaded in or out, health care practices are given, feeds delivered and stored, and it can be the source of water supply systems. If you are fortunate enough to be beginning from

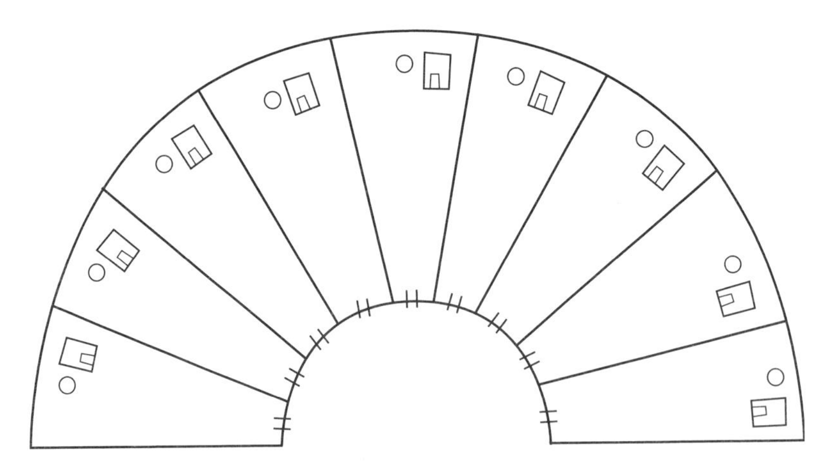

A half wagon wheel layout showing the ease of access from the center hub. This was once a commonly used layout on wooded hill tops throughout the Midwest. A herding dog would be sent in to the pens to bring up the hogs for inspection from the center point of the wheel.

scratch or doing a full rebuild, this is the preferred layout plan. It was even used on a number of university research farms, including the University of Illinois.

Most readers who are getting starting from an existing base will have to deal with pens and lots that grew out rather randomly from the farmyard. These tend to be rough squares or rectangles based on a farm plan with a similiar shape. The classic homestead was a big rectangle; our current farm is a square that, like me, is a bit thick around the middle. The lots and pastures move out from the farmyard, laid out like blocks stacked from one to four cells deep from that central point.

The best penning option here is to have as many of these lots as possible open onto a single shared alley or lane. This service lane, which should be stoutly fenced, provides access to all pens and can be used for some sorting. It should also be extendable for later growth, if need be.

Pie-shaped sections laid out in a circular pattern certainly speed the basic choring process, as all the animals can be drawn to that central point for feed and water. It also provides fast access to all the pens and animals and will facilitate more frequent checks on the animals. This pattern has been used often in the Missouri hill country and can be the most effective way to utilize and access rolling and rocky parcels from a single, easily accessible point. The single access lane in and out of these designs not only reduces maintenance costs, but can also facilitate both regular and biosecurity.

Access gates to pens and lots need to be large enough to assure easy access for tractors and trucks, as well as the movement of portable housing and other equipment. An 8 x 16 foot range house on runners will require a solid, straight-on approach when being pulled into or out of a pasture or lot. These runs also tend to be high-traffic areas for the hogs.

A gate span should be at least 16 feet, set to open both inward and outward, and the posts should be built to last for generations. Dad liked to hang his gates from railroad cross-ties buried at least 3 to 4 feet down and set in a healthy footing of concrete. The other day I passed a gate I built in high school and that Dad had hung over 34 years ago. It was still sitting up tightly and swinging freely. I'm pretty sure that some of those creosoted monoliths Dad erected to hang gates will outlast that stainless steel croquet wicket they have over in St. Louis.

The gateways should be filled with rock to a depth of at least 6 to 8 inches. I have seen water pool at some gates to a depth I believe would sink an 80 hp tractor. Gate openings will need regular maintenance and are not the place to skimp on materials.

To some prospective hog-raisers, it may seem that I am setting the cart before the horse by setting up facilities before acquiring even hog one. But trust me, you don't want to hold one on your lap until you have a safe place to set it down. A hog has pretty simple requirements by anyone's standards, but for optimum performance, they must feel comfortable and secure. And you, on the other hand, will look a lot more professional when they actually stay where you put them.

Years ago, a list made the rounds that noted the different livestock species' favorite methods of eluding containment. Cattle would go over a fence, sheep and goats through any weak spot they could find, chickens under it. And hogs? Well, hogs being fair-minded creatures, they employed all three methods of escape.

I once saw our senior herd boar raise up and shear the stays of a brand new cattle panel with his tusks as matter-of-factly as you or I would trim our fingernails. After turning an $18 panel into so much steel spaghetti, he went over and lay down to admire his handiwork. He was one of those rare hogs that wasn't known for the places he "broke out" (he was content to stay put) so much as for the things he "broke up."

That old boar has long since found his just desserts at the pepperoni factory, but truth be told, most animals that we've owned were not really very difficult to contain. There are numerous fencing options for use with hogs, and some are quite specialized in their design and use. One such type of fencing that isn't seen anymore is a 26 inch high woven wire that was made to be unrolled between rows of corn to contain hogs set to glean among them.

Perimeter fencing is the bulwark of your swine fortress and is especially crucial in these litigious times. Its task is to keep them off the road, out of your neighbor's pool, and most importantly, out of your wife's flowerbeds. If it also discourages predators, deters rustlers, and looks nice out on the road, then that's so much the better. Possibly the very best choice for a perimeter fence for hogs is a combination of woven and barbed wire.

Begin your perimeter fenceline with a strand of four-barb wire suspended roughly 2 inches above the ground. It should be clipped to steel T-posts set somewhere between 8 and 12 feet apart. Top that strand with woven wire that is at least 32 inches high. Four and 8 inches above the woven wire goes two more strands of four-point barbed wire. This is about as close to a "hog-proof" fence that we as mortal human beings will be able to create.

Occasionally, a good buy can be found in new or used hog panels, which are 34 inches high and 16 feet long. Use them to replace the woven wire in the above configuration and attach them to steel posts on 8 foot centers. This should be one of the most durable of all fencing choices, and it's also one of the easier to take down and/or reconfigure.

The old rule of thumb (sometimes even encoded into state laws) is that a perimeter fence between two farms is the equal responsibility of both landowners. Still, an owner cannot force an adjoining landowner to help erect a fence if he's the only one with livestock to keep in. By a sort of gentlemen's agreement, these matters are usually settled, and I have never seen the law used to force someone to comply. Usually, a compromise is made, like substituting your added labor for actual materials. One way to execute this process is to have the two landowners meet at the exact center of their shared property line, and then the fenceline to each person's right is his or her responsibility.

Another point to consider when creating perimeter fences concerns health management. Nose-to-nose contact is one of the surest paths for contagion to spread. A buffer of at least 8 feet has been found to be adequate to curtail the airborne spread of disease-causing organisms. To that end, you may well use a second, often electrified, fence to keep the animals back from groups of the same species contained on adjoining farms.

Interior Fencing

Interior fencing is typically employed to serve a twofold task. It is used to keep hogs both in and out of various places on the farm. Cropland, waterways and impoundments, high-traffic lanes, the farmstead, and many other sites are fenced to keep hogs out—out of mischief if nothing else. These lines, like those for permanent pastures, should be tightly fenced with the most durable fencing materials available.

Those 16 foot long hog and cattle panels have changed the face of livestock farm fencing, as have few other equipment developments. A pickup bed of these plus a few bales of steel posts, and hog lots can go

see page 52

FENCING

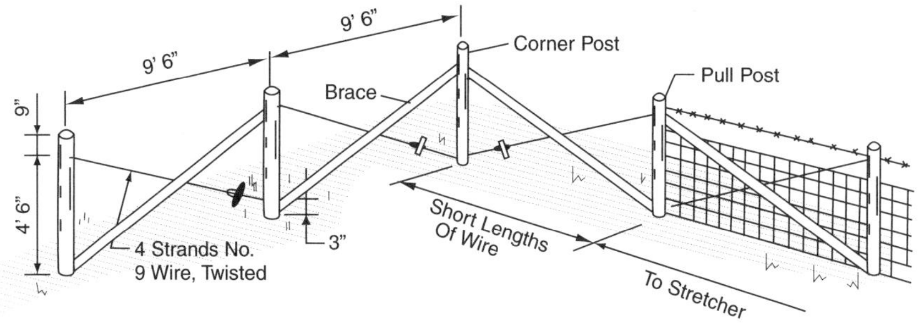

Corner or End — extra strong

Construction Steps

This extra strong fence corner, or end, is good in soft soils, or where deadman or corner post would otherwise be necessary.

1. Set all fence posts.
2. Install bracing.
3. Fasten wire to second post.
4. Tighten from second post, and complete line fence.
5. Using short lengths of wire, close corner.

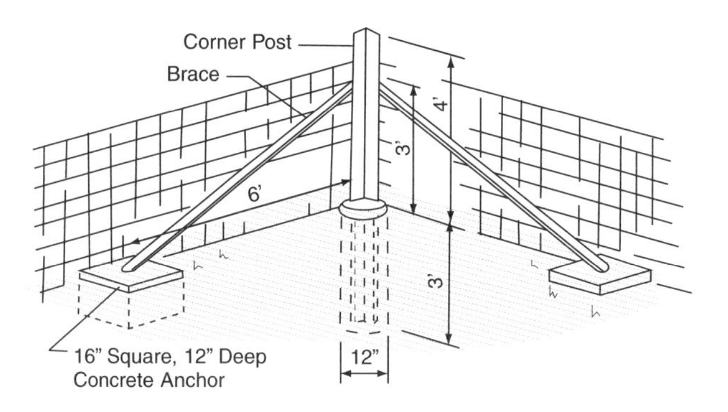

Corner or End — steel strong

Post Sizes

End Post, minimum sizes

$2^{1}/_{2}$" x $2^{1}/_{2}$" x $^{1}/_{4}$" Angle

2" I.D. Standard Pipe

5" Top Wood Post — 8' long

Brace Post, minimum size (wood)

1st Brace Post — 5" top, 8' long

2nd Brace Post — 4" top, 8' long

Brace

For Angle or Pipe Corner Post

$1^{1}/_{4}$" I.D. Standard Pipe

2" x 2" x $^{1}/_{4}$" or $^{3}/_{16}$" Angle

For Wood Corner Posts

2" I.D. Standard Pipe

2" x 2" x $^{1}/_{4}$" or $^{3}/_{16}$" Angle

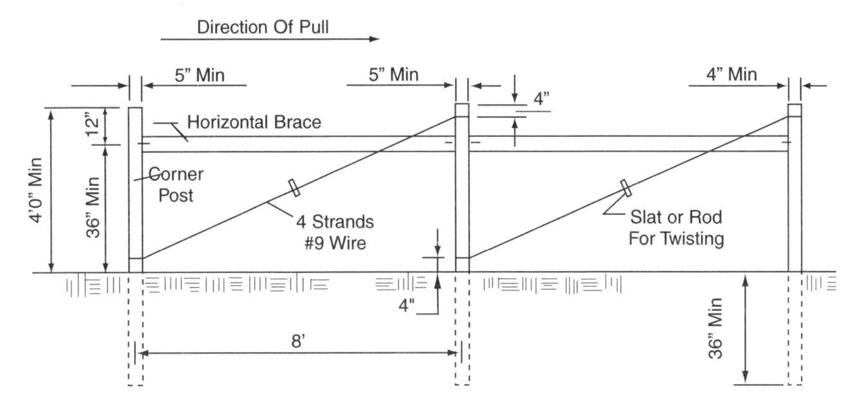

Corner or End — wood posts

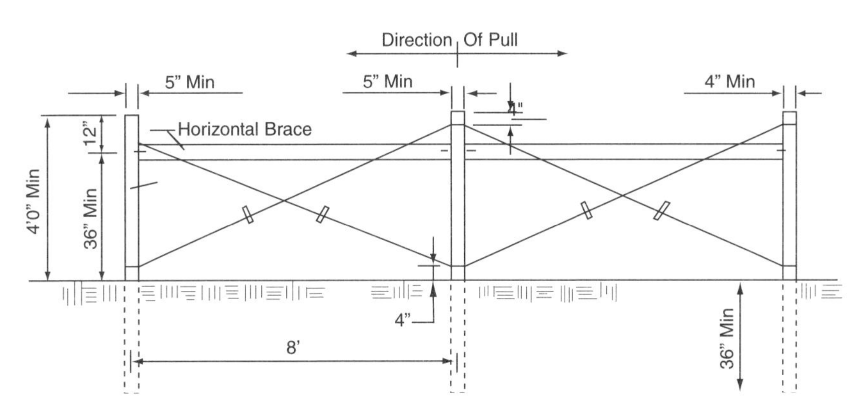

Wood Pull Post
For middle of long fence, place about 6' apart.

Braces

3/8" x 3"
Lag Screw

2" Pipe

Notch Post
3/8" x 4"
Steel Dowel

3" Diameter Pipe

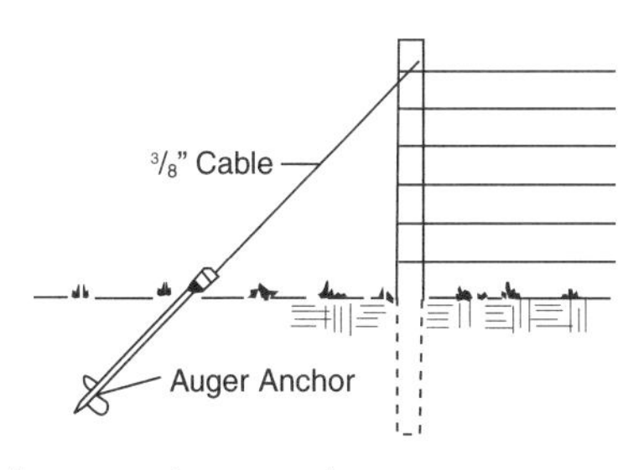

Auger-anchors can brace corners and ends.

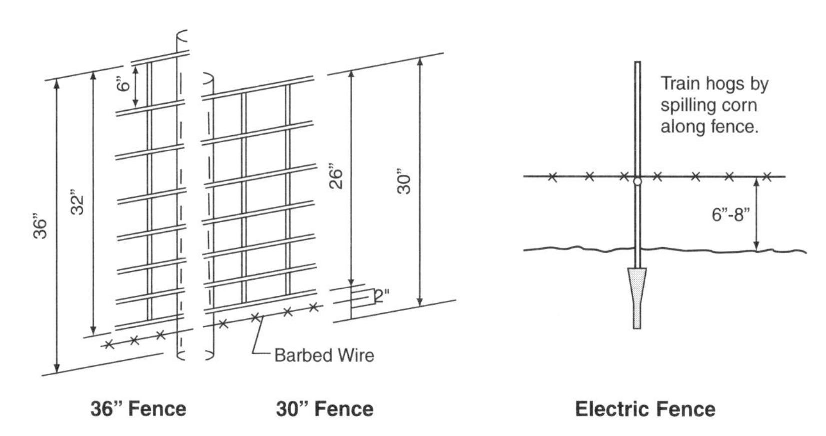

30" Fence: Can climb over.
High enough for temporary fence and between pens where pigs can see each other.

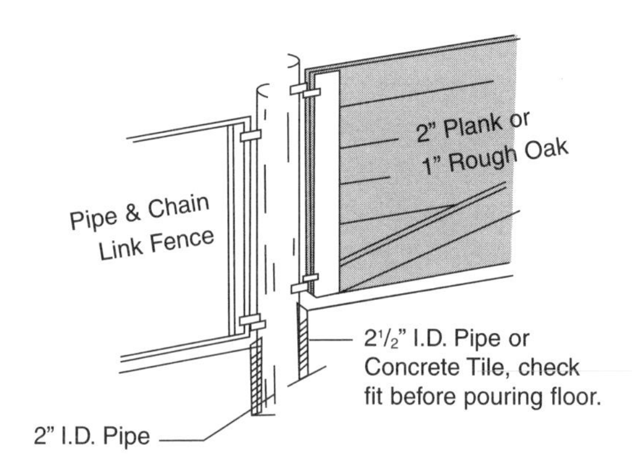

16' Hinged Panels

GATES

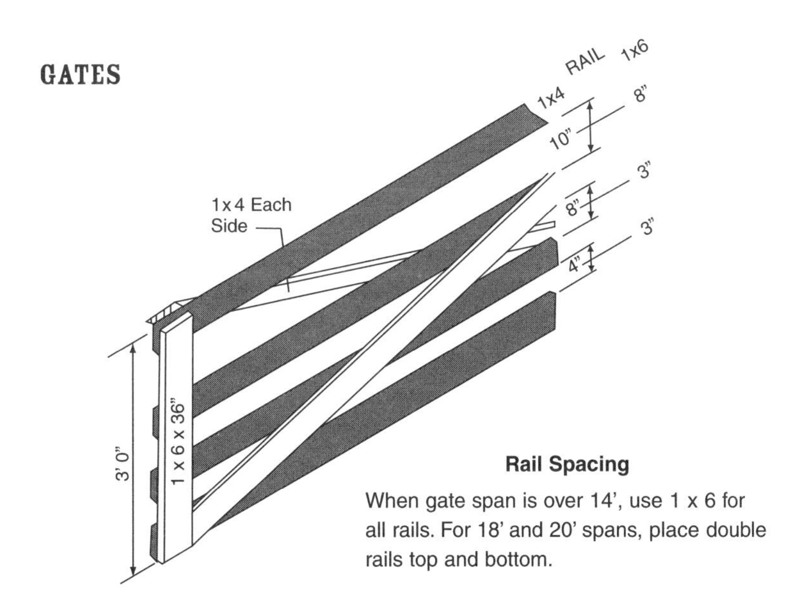

Rail Spacing

When gate span is over 14', use 1 x 6 for all rails. For 18' and 20' spans, place double rails top and bottom.

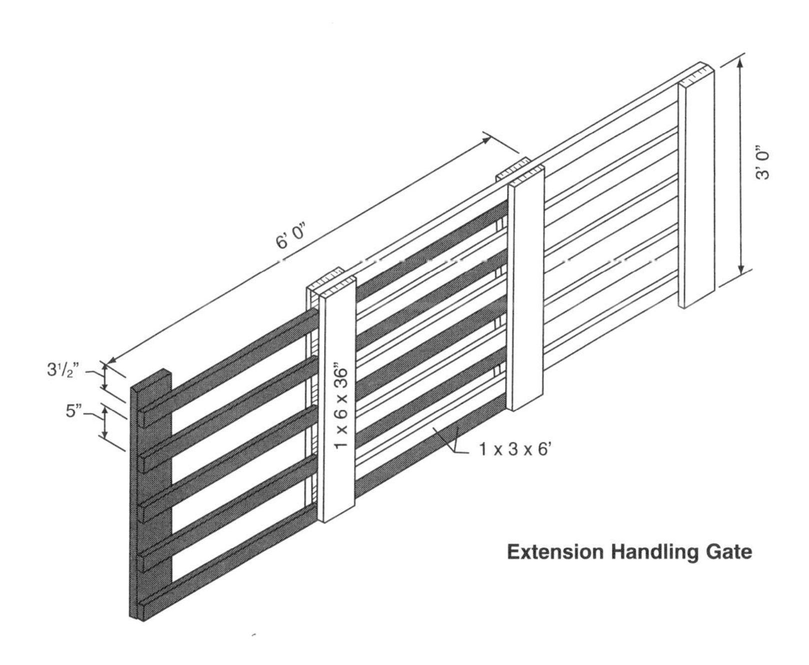

Extension Handling Gate

HINGES

Note: Install hinges on both ends of gate for double hung partitions.

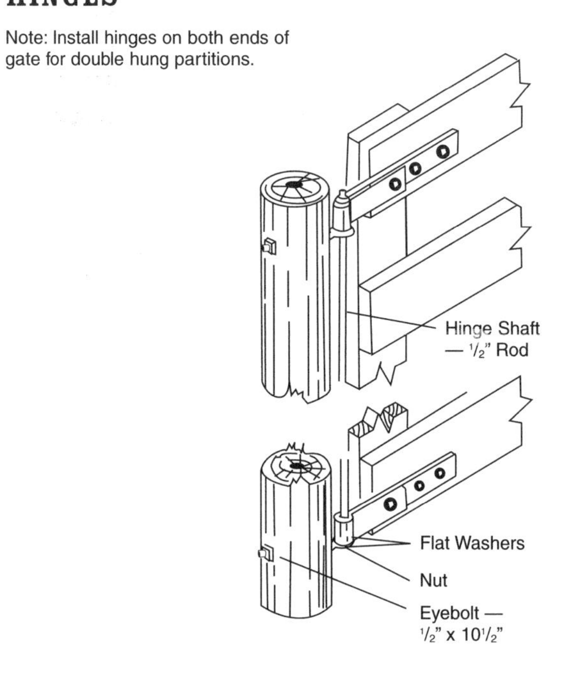

Lot & Field Hinge

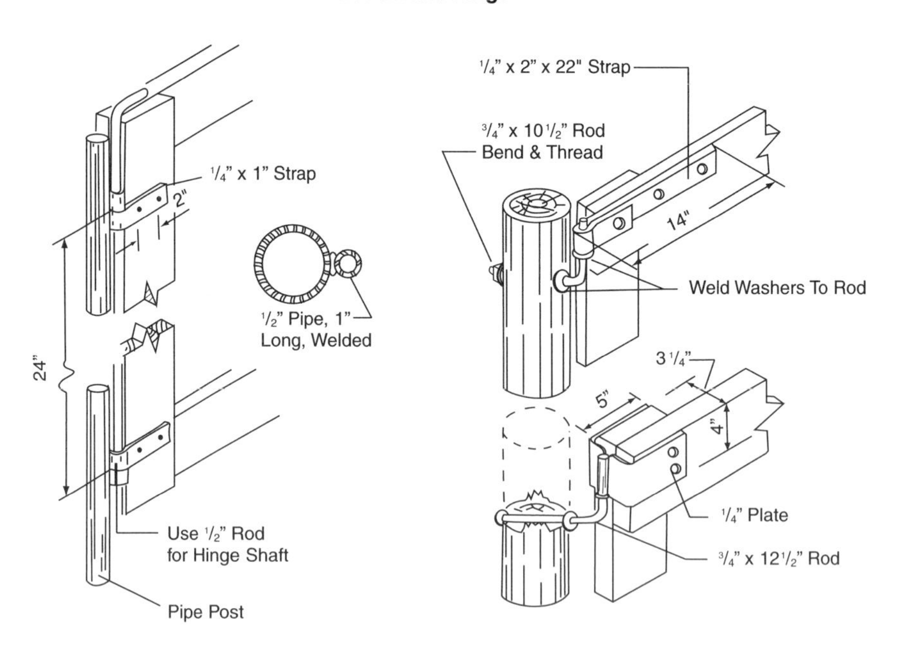

Pen Hinge

SOW WORK AREA ALLEY

Pole Building — 18' x 48'

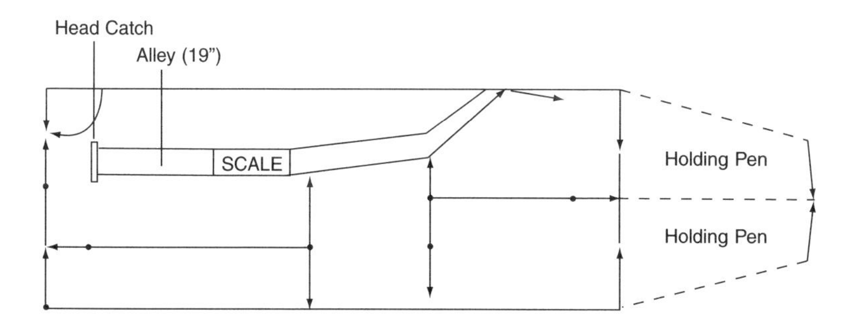

CLASSIC RANGE HOG HOUSE

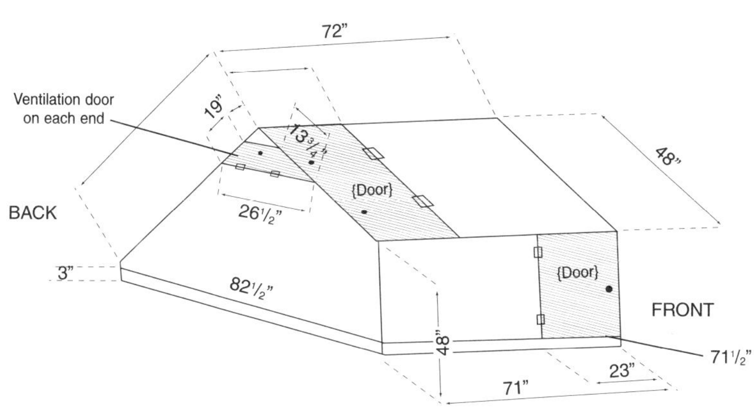

1 1/2" overhang on top of roof and at bottom of back roof. Two boards on back roof (One 4' wide and the other 20" wide).

Overlap of plywood over 2" x 6" about 2 1/2"

from page 45

up in hours instead of days. Over the years, they have gotten markedly cheaper, and lately I have been able to buy many used 34 inch high hog panels for $3 to $5. Real buys over the years have been stacks of these panels cut into short lengths and then taken down and sold after they were no longer needed for their original use (generally in water gaps). These short lengths often prove as handy as a shirt pocket.

The interior fencing choice most favored by range producers is electric fencing. Over the years, "hot wire" has gotten consistently better, and it can be used to contain hogs of every size and age. Farmers with early bad experiences with electric fencing would be well advised to give it a second look.

Early electric fence chargers were a bit touchy to use, shorted out quite easily, and were rather limited in how they could be set up and used. Most 110 volt models had to be set up inside a building, and the battery-powered models were not nearly as efficient as today's. One of our early chargers had a plug-in chopper-controller as a sort of surge protection device. A crack of thunder within a hundred miles of home, and we were offline until another one of those little $5 gems could be plugged into the charger body.

Over time, batteries have gotten better and more dependable (one charger unit now even uses flashlight batteries), solid-state systems now make the 110 volt systems far more dependable, new case designs make them more flexible in their use, and a solar option has been added to expand electrical fencing's operational range.

The controller is the core of the electric fencing system, available in sizes adequate to charge 2 or 20 miles of fencing. Many models can now attach directly to the fenceline and remain there in all types of weather. Most of the rest need little more than the shelter of a three-sided box to keep the moisture off of them. Solar models can be set up far beyond the nearest electrical outlet, and some of them can be combined with a backup battery to assure dependable service through extended periods of cloudiness.

A number of firms now design and market electric fence controllers, drawing together design elements and materials from all over the world. A design group currently doing well now is that of the "New Zealand" fencers. They take their name from that island nation where they were developed to contain sheep and cattle over vast and rather remote holdings, generally in the presence of predators.

These fence systems are well made, quite powerful, and are capable of powering up many miles of fencing. They aren't cheap, by industry standards, but they have proven their worth, containing one of the most difficult livestock varieties to keep in with electric fencing, wool sheep.

Basically, an electrical fencing system breaks down into four fairly simple components. They are the charger, the wire, insulators, and posts. All four exist in great variety, and there are a few additional ruffles and flourishes, but those four elements constitute the basic fencing system.

There are $29.95 fence chargers out there, and others that will crowd the $1000 mark. Some will charge 1 to 3 miles of wire, and others will loop seemingly endless sections of land. Which one is the best?

We have a neighbor that is actually charging two strands of barbed wire with the flashing element from an old neon sign. It is well respected by all of his stock, and with very good reason: the pulsating 220-volt charge it carries will, as my grandfather used to say, "knock you a-windin."

The selection process begins by assessing your actual needs. Do you need to charge a few strands of dividing wire or multiple strands of wire spanning acres and holding all manner of boars and sows? Will you be fighting a lot of grass and brush along fencelines, or can they be regularly and easily trimmed? What about your winter weather pattern? Lots of snow can build up in drifts over the lower strands of a charged line, shorting them out. Insulators are available that can be slid up and down the post to raise the wire away from such an obstruction.

I was taught that while it is possible to do small jobs with big tools, the reverse is simply not possible. Buy the best unit with the largest capacity that you can comfortably afford. Also, bear in mind that these units now have a much longer lifespan, so be sure to factor in future fencing capacity needs.

The 110 volt models are generally the best choices where electrical service is easily accessed. A friend of mine has one in a barn that has run day and night for the better part of a decade. The charged wire passes through a small hole that has been drilled into the barn wall and insulated with a bit of scrap garden hose.

There are weatherproof models that can be mounted onto a fenceline and powered up through a good, all-weather extension cord. Still, you should use a bit of extra care and cover it with a simple shelter of some sort. If nothing else, upend the bottom part of an old Styrofoam cooler atop it.

It is also good to bear in mind that 100 feet is about the maximum length for an extension cord supplying such a unit.

An electrical service point can be established at any utility pole adjacent to the desired fenceline. We had a meter base set on one such post about a half-mile from the house. At its base, I set an old wooden shipping crate atop a couple of concrete blocks, caulked all the seams in it shut, slapped on a couple of coats of paint, and wired in a second-hand fuse box and then an all-weather outlet. It became a shielded place to hang a charger and even store a few small items.

The battery-powered models generally come with room in the unit's case for a dry-cell battery. It is watertight and enables the charger unit to be hung adjacent to any fenceline. Such units generally sell in the $50 to $75 range, and the battery will sell in the neighborhood of $20. The batteries normally last from 2 to 4 months, but of course their charge will tend to weaken over time. What's more, they are not widely available and generally aren't up to the task of larger fencing jobs. Also, you have to remember never to store a charger away with the battery still connected and in the case.

These units have also gotten better over time. Some will take larger batteries, and a few now come with solar cells to help the batteries maintain a strong charge. Some of these units have been modified with longer leads to accept wet-cell batteries, including the far longer lasting deep-cycle models. Such batteries can be placed beneath the charger on a block or wooden riser. The batteries should also be positioned so that the animals cannot have any contact with them.

Battery-charged lines should be checked often using inexpensive testing units that can give a quick reading on the charge's shock strength in the line. It is always a good idea to have a spare fresh battery on hand.

Numerous solar chargers are available, and good ones can be had in the $200 range. Many come with a backup battery for extended periods of overcast weather. They are much sturdier in design than earlier models and can be set directly on the fenceline. They perform best when fencelines are kept clean and will benefit from frequent inspections. These may be the best choice for distant installations.

On many range operations, you will actually find a combination of charger units in use. We have backed up a 110 volt unit with battery-powered models. We now have one of the units that use flashlight batteries for small and short-term uses. Hot wire has been a staple on hog farms for over 4 decades, and the technology just keeps improving.

We've used it to contain 2-week-old pigs and 600 pound boars. It is not foolproof, and we once lost a boar when it fell beneath a charged strand on glare ice. Ultimately, the system's setup, operation, and choice of fencing materials are all key to its success.

Electric Fencing Wire

Electric fencing wire comes in a variety of gauges and can be made from a variety of materials. There are woven types in which very fine alloy strands are interlaced with synthetic webbing. There are also similar varieties that look like nylon rope, and some that come in elaborate bars, strands, and strips made from a combination of synthetics and charged wire strands.

Smooth wire is still probably the most common choice for swine containment. There are some alloys with a high aluminum content that are very light and modestly priced, but they aren't very durable and can quickly become brittle. The heavier the gauge of the wire, the less conductive it is apt to be. Very light wire, on the other hand, is prone to kinking during handing and can easily be broken. Perhaps the best advice is to match your wire selections with the recommendations of the charger manufacturer. It's not uncommon these days to find chargers, wire, posts, and insulators all offered by the same company, tailor-made to be used together.

Setting Up the Charge Box

Just as location is the key to successful real estate sales, the installation of the charger is equally crucial to a successful fencing operation.

I like to set the controller someplace that it will be easily accessible and can be given a quick scan several times a day as I go about my chores. On the homestead, our charger hung for many years in the granary at roughly shoulder height. We passed the charged wire around a corner wall in a piece of conduit made from simple PVC pipe. The first thing I looked for every morning and the last every night was the flashing light on that old controller, which assured that it was delivering a solid, short-free charge. It also produced a reassuring clunking sound as it operated.

When positioning a controller along a fenceline, securely mount it on a wooden or steel post set specifically for that purpose. Again, it should be in an easily accessible and highly visible position. It should also be far enough away to prevent any possible contact by the animals. A simple mantle-type hanger box can be made from treated lumber to be used both to attach the charger to a post and protect it from the elements.

To operate correctly, the charger must be well grounded and the grounding rod always in contact with damp earth. Most farm stores now sell 6 to 8 foot copper rods about as big around as your thumb that are ideal for this purpose. The deeper you can pound them in, the better, and in some very dry climates, you may have to sink them a full 6 feet. In extremely dry years, you may even have to dampen the ground around the rod to ensure proper charger operation.

Use good quality, smooth wire to link the charger to the ground rod. I like a simple loop pulled down tight using a wing nut on the charger's grounding post. I then wrap the wire tightly around the grounding rod in several loops to create a wrapping

But not everyone has a hardware store on hand when you need one. I will admit to having had moments of desperation in which I spliced baling wire into a fenceline and left it to chance that it would do the trick. Still, since electric fencing is frequently taken down and put back up as lots are rotated or restructured, it really isn't a place you want to skimp on quality. Years ago, we found ourselves heirs to several rolls of old telephone wire. It was stout enough for a buffalo to roost on, would hollow out the jaws of the best bolt cutters in short order, and despite this, it got sent to the local sale barn just as soon as we could get it rolled back up. When it came to lighter electric charges, it just wasn't up to snuff.

Also available are a few types of lightweight, two-barb wire for use with electric fencers. I've found, however, that they aren't easy to handle and don't really contain hogs any better than smooth wire.

Now, I know you're going to say, "Kelly, I have neighbors using multiple charged strands of conventional, four-point barb wire." I have one who does that, too. That old heavy-gauge wire has poor conductive qualities, doesn't really fit any type of insulator I know of, and at least 1 inch high. Over these wraps goes a good brass clamp turned tightly to assure solid and lasting contact. Inspect these contact points often for any signs of loosening and corrosion.

When rushed, I've run the charger lead to the fenceline and positioned it with little more than a few short wraps on the fence wire I wanted to charge—not a really a good idea.

Ideally, such wraps should then be clamped solidly into position, or better, the leads can be coupled to the charged wires with alligator clips crimped snugly into place. Short loops of stiff wire with alligator clamps on each end can be used to deliver the charge to multiple strands in the fenceline. Short bits of wire with alligator clips on the ends are also useful in transferring the charge at corners and from lot to lot.

The controller is not a "set it and forget it" kind of appliance. It will need regular inspection and should be easy to access for servicing. But it's worth the trouble. Being able to step out on the back porch on an evening to see those red charger lights pulsating, attesting that they were doing their job, certainly made sleeping easier at night.

doesn't generate the "piranha bite" many believe it to have. I buy the brand-name wire in 1 mile and half-mile rolls, and for the few dollars extra, I get wire with far longer life and easier handling qualities.

Posts & Insulators

These two components are probably best discussed together, as they have to work well in unison. Due to the power and forgiving nature of today's charger units—they are now more durable and simpler to operate—all sorts of post and insulator options are now available. Still, good old farm ingenuity has been known to suffice here, too. More that once, I've seen an old glass pop bottle up-ended on a piece of rebar, with the fencing strand looped around it. Of course, this can leave you with a bit of an image problem with the buying public—but, hey, nobody said recycling had to be pretty.

There are insulator varieties out there that make it possible to adapt either wooden posts or steel T-bars for use with electric fencing. There are also a number of fiber and metal posts available now specifically made for use with electric fencing. Some even have specially engineered step-in or push-in features for rapid installation. The fiber models often have built-in insulators or else may be predrilled to ease the fastening of wires or clips. A few of the early models would become brittle in cold weather and would easily snap off, but that no longer seems to be a problem.

There are even a few systems now that combine wire mesh fencing and factory-installed step-in posts all in one package. The easier they are to pull and use again, the more they'll be in keeping with your needs as a range producer. Experience quickly teaches us that the best choice is what you already have at hand, but if posts must be bought, it is hard to beat steel T-posts in the 5 to 6 foot range for greatest versatility and durability.

The role of the insulator is to hold the wire strands at the desired heights and without being ground out on the post surfaces. These days, there are a huge variety of insulators, from that old classic porcelain donut to plastic varieties that can mount charged strands to existing fences to keep animals from probing them for any weaknesses. Virtually all of the first generation of insulators were made of porcelain, and the great question is, where were ATVs when we were packing buckets full of those things up and down Missouri hillside pastures.

There are now plastic insulators that can be clipped or clamped to posts and slid up or down, and still others that can be nailed into existing wooden posts. Wooden post insulators generally come 50 or so to a bag and cost 20¢ each or less. We live in an area that will get at least one or two good snows each winter, so we favor insulators that are easily adjustable up and down the line posts. A twist of the wrist and you can move them a few inches higher ahead of any expected snowfall.

Those old porcelain donuts may still be they best choice for corner post insulators, and they seem to inspire all sorts of creative uses. They can be used to suspend charged wire in any manner of configuration, they are the most durable and are as inexpensive as they come.

Also available now are a number of clip-action insulators that can be opened to accept the wire and then will close around it in some fashion. If you prefer to go this route, shop carefully, to be sure that you're getting both ease of use and durability in a clip. I once bought a set of these that would confound an engineer in their use, and they still work their way to the top of my insulator bucket from time to time. State-of-the-art clips and gadgets that are too complex for practical use can overlap and conflict on occasion, so you should beware of any overly complex designs. The more complex they are, the more likely they are to ice up or collect water and cause a short. Cracked or misaligned insulators can also cause charged lines to short out.

I try to keep a few of most kinds of insulators on hand for emergency replacements anywhere along our fences. I buy used insulators and posts at farm auctions when they can be had at a good price. They are in that group of supplies that seems to always need replenishing.

Setting Up Electric Fencing

My dad held that the best farm fencing was horse-high, bull-horn stout, and baby-chick tight. Of course, such a fence now would set you back slightly less than a hundred bucks a running foot.

Perhaps this is the time to state the obvious about electric fencing: It cannot actually restrain anything. It is a psychological barrier that functions because of the "pain principle." Touch it and it will sting you. But if you push the animals too hard or get 'em scared, they will go through it. There are sows out there, and no one knows how they do it, but they can tell the when a charger fails.

If they've learned to respect the fence, hogs on range can be contained with a single, well-maintained strand of hot wire. For maximum containment of hogs of various sizes, an arrangement of three charged strands is better. They should be strung at 4, 12, and 18 to 24

Two charged strands of electric fencing containing a good set of crossbred gilts. This may be the toughest age group to contain, and this simple, two-strand system does the job well and inexpensively.

inches above the ground. Charged strands at 4 and 12 inches are generally the pattern of choice for sows with nursing pigs. With these heights, the hogs will first encounter the charged wire with their face and nose, places where the shock will have maximum effect.

When using charged wire to define pastures and lots, line posts can be safely positioned 16 to 20 feet apart. Old-timers don't like to see the line strung fiddle-string tight either. It is less likely to break if pressured, and some believe that the shock effect will actually be magnified if it drapes slightly across the animal.

Whenever possible, I like to lay out my lots and pens with four distinct corners or at least in configurations with straight sides that are as long as possible. At each corner or starting and stopping point for a straight run, I will set a 6½ foot T-post for a corner post. My fencing then becomes little more than a series of straight pulls from one strong post to another. Moving from one deeply set post to another, the pen is soon enclosed with fencelines that are easily set and maintained.

We use porcelain donuts for the corner insulators, and it is very easy to draw the wire straight through them. We then tie the lines together at each corner using short strands of tie wire coiled tightly around the long runs, or else clip to them using alligator clips crimped to each end of the coupling wires, as mentioned above.

With a bit of care in planning the layout, we are able to power several pens and lots with a single larger controller. You can link them together in a circuit of sorts, but maintenance is easier if they can be separately disconnected from the "circuit." To do this, essentially create each pen as a separate unit and then tie them into the loop with short lengths of wire coupling. Another approach is to wire simple, blade-type switches into the line at the head of each pen. Then a simple flip of the switch will take a pen offline for repair or reconfiguration.

Hogs quickly come to respect the wire, and some will get skittish about crossing a place where a charged strand was once hung. A variety of spring-type "gates" are sold for electric fencing, but I have found very few of them to be truly durable or easy to use with hogs. Also, sorting at gate areas or driving hogs in or out can be quite difficult with some electric fence configurations.

To facilitate those sorts of movement, many farmers use panels or heavy gates for a part of the fenceline adjacent to the entry and exit points. A pen within a pen can thus be formed by arranging panels or gates for sorting, unloading, or containment for any type of medical treatment. You should first feed them within that mini-pen for a couple of days to set them at ease in it before enclosing them in the larger pen. Erect these pens with posts set on 8 foot centers to keep them solidly in place in case hogs are crowded during sorting or have to be restrained.

A few companies now make drive-through gates for electric fencing that enable you to enter a pen with a truck or tractor by pushing through the overlapping gate arms. They will not work for young hogs, however.

With long runs of electric fencing, it may be necessary to pass lines over roadways, gates, or low places where water can accumulate. All are tricky, but they can be overcome with simple variations of two themes. The first is to elevate the line on poles high enough to drive under or to allow water to flow well below it. The alternative is to route it through some sort of conduit and run it underground.

Whenever possible, run your lines across open terrain. Keep it out from under trees, where falling limbs can short it out with every storm or strong gust of wind. The wire should be strung on the inside of the posts, as an added strength factor. With just one or two strands and on open terrain, it may be possible to set the line posts at a sufficient angle to facilitate mowing beneath the wire from outside the pen. Mowing helps to reduce shorts and current drops caused by plant buildup along the fenceline.

A high-quality testing tool is important to have around to frequently measure the strength of the charge in the fencelines. Inspect the lines after every storm, and outfit a 5 gallon pail or a canvas tote with tools and supplies to carry along as you make your inspections.

Electric fencing has certainly come a long, long way since we started using it nearly 40 years ago. Some of those early chargers were near-

Breaking Hogs to Hot Wire

On first exposure to charged wire, most hogs will simply turn tail and run straight through it. They don't yet know enough to respect it.

The first contact with electric fencing is best accomplished in a pen enclosed with a physical barrier type of fencing. We generally do this in a barnlot or other simple holding pen. There, it is a simple matter to run a charged strand through a portion of the pen and allow them to gather in there more or less on their own terms. String a bit of feed close to it to be sure they quickly get its electric message.

In the first few days on range under a hot wire, the hogs need to be closely monitored. And as with all types of fencing, some escapes are to be expected. Sometimes they will simply begin running and playing and pop through, or else dogs or other predators can set them running through a hot fence. Hogs will also find most fallen branches on fencelines well before you do, though to get them back into the pen, you will probably have to drive them all the way around to a non-electric gate.

Never turn a hog that's new to the herd right out into an electrified fencing enclosure. A reordering of the pecking order will generate running and fighting activity, and hogs will be driven through the wire. The electric charge also seems to have a stronger effect on intact males than on any other sort of hog. Put them out only when they've learned to respect the wire, however.

ly the size of a good bale of straw. They were temperamental to use and weren't nearly as durable as they are now.

Still, you cannot simply turn them on and just forget about them. Hot wire will not keep adult boars apart, nor will it keep freshly weaned sows away from their pigs, if they are held nearby. And versatile though the fencing is, it's just one of many kinds of fencing you will need in an outdoor operation.

It should also be clearly stated that no matter what the salesmen may say, there is no such thing as "hog proof" fencing. Escapes will happen, and some, you'll find, are something akin to porcine magic. They'll be inside the pen one instant and ambling through the front yard the next. Hogs are no more prone to escape than any other livestock species, however, and thankfully they are not terribly inclined to roam, should they get out.

The most common escape routes are around gates, at water gaps, where fencelines dip sharply, and in areas of high traffic. Upon first entering a pen, hogs also are likely to work the perimeter, looking for weak points. All farm fencing should be walked regularly, and potential weak points mended as soon as they are detected. Be sure to inspect all the fencelines after every storm.

Genetic progress for specific goals did not—could not—begin until animals would stay where they were put. They are not called controlled matings without reason.

In range or outdoor production, you will be moving the animals through a number of lots and pens, often from one farm to another. Something to be valued then in fencing options is portability and ease of setup and operation. A time or two in my life, I have actually built paneled pens around hogs "in motion." Fun it ain't. Invest in quality materials, handle them correctly, and keep them well maintained, and be sure to keep a detailed pasture and fencing map. It is far easier to move fencelines around on paper first. Make it a part of each New Year's resolution to pull out the maps and records, to give everything a good review, and plan for changes in the upcoming year.

Housing Basics

Taking hogs outside doesn't mean you've eliminated the need for roofs and walls. They still require some sort of shelter that will go with them.

I once knew a Missouri producer who did eliminate the need for roofs and walls for his spring-farrowed hogs each year. When the hogs were weaned, he would move gilts and barrows to one pen and keeper boars to another.

These "pens" were the gravelly hillsides for which much of eastern Missouri is noted. High up on each hillside, he would drop two or three large trees in a crisscross pattern. Thus, he created a leafy bower or arbor that would shelter the growing hogs while giving them the fullest possible open-air experiences.

I've seen piggy sows unloaded one-by-one among large bales of fescue or other common grass hay. On the flat ground, these bales were quite stable, and the sows were able to burrow into them to create snug, dry nests for range farrowing. Not only were the sows able to experience great freedom on range, but they were doing it with housing that was not only low-cost, but was also biodegradable.

Successful hog housing will keep the animals dry, protected from drafts, and up out of the muck. Nearly all range housing is what is termed "cold housing:" the temperature inside moves up and down with the outside temperature. It does, however, create a microclimate that

A one-sow house set up in a machine shed end rigged with an inexpensive box fan for hot- weather farrowing. It is set to blow through the pig bunk and is thus protected from the sow. Such a measure is seldom needed if farrowing is planned around the extremes of weather, but it does illustrate the creativity of a good range producer.

protects the animals from direct sunlight and holds in body heat in the winter. Four to 6 inches of clean, dry bedding will increase winter comfort levels by several degrees, as well.

There are pork palaces and swine tenements on the nation's farms. Looks can be deceiving, however, but as long as they are providing these few protections, they are doing all that can and should be expected of them.

Outdoor hog rearing is essentially pork production in motion, and the housing used must have the ability to be packed up and move along, sometimes quickly. I once bought two large double-farrowing huts with floored pens from an Amish carpenter, and they arrived stacked quite neatly on a single flatbed truck. Most range housing is built on 4 or 6 inch wide treated runners, and many a hog house has been pulled from one farm to another behind a tractor or pickup. They're called "early Sunday morning trips," since that's the best time to avoid traffic and patrolling deputies.

In a range operation, you will need housing for the breeding herd, for growing hogs and the sows at farrowing. When well built, this type of housing easily has a lifespan of 10 years. For tax purposes, they can be depreciated out quicker than this, and with slight modification, much of it is usable with other kinds of livestock. It should be built to meet the task, but no larger.

These are low-profile, modified A-frame farrowing houses. Your great-grandfather used houses like these, although made with tongue-and-groove lumber rather than plywood. The openings in the end are for summer ventilation, and they should he painted white.

There are runnered houses out there now that are crowding 40 feet in length and 16 feet in width. They are movable, but only with tractors cranking 100-plus horsepower. They also tow and corner much like the aircraft carriers that they rather closely resemble. If considering buying such housing, first envision getting it through the gates back home in the muddy season before you make an offer. If it takes permits and teamsters to get it home, you probably don't need it.

To stand up to the wear and tear from hogs and the stresses and jolts from frequent movement, they have to be well built, carefully handled, and always put to use after plenty of forethought for their placement. When we build swine housing, we prefer to use full-cut, native lumber for both the framing and the runners. Here in Missouri, that generally means white oak lumber from local sawmills. Sadly, good 4 X 8 runners much longer than 14 feet now are very hard to find.

For ease of construction and to accommodate both corrugated sheet tin and plywood, we like to build everything on 2 foot centers. You learn early on how to swing a heavy hammer and to get that native oak nailed down while still green. Let it dry too much and it will burn up drill bits trying to put holes in. In framing hog houses, we like to use the biggest ring-shanked nails we can drive to get a good, lasting hold, since the frames flex when moved.

Many folks may not have access to this kind of lumber, but other types of lumber and construction elements will also work. Treated lumber is widely available and is quite serviceable. Treated runners don't seem to last as long in very wet situations, however. Also, if using treated lumber, be aware that some types of treatment can cause skin problems with hogs. This is especially true of White hogs or hogs that lie on large areas of treated material. Recycled steel pipe is now being used to frame up some range houses and even serve as runners, especially for bigger houses.

The Range House

Outdoor hog production actually lent its name to a type of housing that has been in use for generations. The "range house" has been used to shelter growing hogs, breeding animals, sows and pigs, and weanlings for ages. These low-roofed, runnered buildings dot pastures and lots from Texas to the Canadian border.

They are three-sided buildings, the original design of which has one of the two long sides left open. I have seen them built in lengths from 8 to 24 feet and widths of 6 to 12 feet. Some of the bigger units were built on permanent pole or post construction. The runnered units generally top out at about 8 feet wide and 16 feet long. Runners much longer than that are rather costly and are to portable housing what Gabby Hayes is to modern Chinese cinema.

These houses are generally positioned with the open side facing south, away from prevailing winds. This is to keep drafts and blowing snow or rain off of hogs bedded there. An 18 to 24 inch overhang in front also helps to prevent rain from blowing into the bedded areas.

Range houses are generally made with simple, shed-type roofs that are 5 to 6 feet high on the open front side and drop to 3 or 4 feet high on the back wall. These are built atop two 64 x 4 x 8 inch runners positioned down the long axis of the building. The longer the building, the bigger the runners should be, as they provide much of the rigidity needed for safe transport.

Tin roofs should be insulated with blackboard insulation in at least 1/2 inch thick sheets. This or other similar durable insulation material keeps body heat in and prevents moisture from condensing on the cold underside of the metal roof and falling back on the hogs and the bedding. The blackboard insulation has proven to be very durable around hogs. Other types of material and sidewall insulation can be

Hog House Runners

Runners are essentially the foundation of a portable hog house, and as such, are the key elements to its durability. The house will last only as long as they do.

The runners should be beveled on each end, and it should be possible to pull the hog house from either end. Tow points should be positioned at least 1 foot in on the runners, to prevent tearing them out in a hard pull. From each end, go back at least 1 foot and bore a 1 inch hole in the center of the runner. Through this hole insert a 1 inch bolt at least 4 inches longer than the width of the runner. These bolts (two per end) become the anchor points for towing chains or straps. Better than the bolts are U-shaped straps bolted in at the same point and extending a couple of inches beyond each runner. These should be made of at least 1 inch wide metal.

The runners should be inspected often and replaced if they become excessively worn or damaged.

protected from the hogs by applying 1 by 2 inch or other wire mesh in places of high activity.

Side and back-end walls should be completely enclosed with sheet metal or plywood. A lot of old hog houses were once enclosed with tongue-and-groove lumber, but this option is no longer cost effective, and the other two methods are less prone to failure. The objective here is to create completely draft-free containment from the three sides of the house that are turned against the wind.

Corners are especially crucial elements of range house construction. They are the "make or break" element, the point where most range houses are apt to fail. They often pull apart at the corners during towing, if not properly braced there. Corner studs should be braced from both sides and at both the top and bottom. The same is true for the center studs on both sidewalls. The more bracing that can be built into a corner, the better.

Range houses are sometimes called sleeping sheds, since this is indeed their main function. They may offer some secondary shade, but for hogs on range, their primary purpose is to provide a warm, dry place to sleep.

Moving & Setting a Range House

Several years ago, we went literally overnight from a wet and muddy fall to a hard, cold winter. One of the consequences of this was that we were left with several range houses frozen into the ground.

One of the shelters we were unable to dig free, and was thus out of use for several months. Freeing the others was a test of both body and mind and cost us several dollars for pain balm.

The basic guidelines for range house handling include:

1. Never position a range house inside a pen. Rather, pull it adjacent to or just to the fenceline. This way, it can be pulled across ground that hasn't been churned up by the hogs.

2. Start it moving with a slow and steady pull and absolutely no jerking. A jerk might free it up if held down only so slightly, but it can also tear out the end.

3. When you get it going, pull the house about slowly. All turns must be wide and gradual, and your route of travel should be well thought out. I was once told that the best way to move these was to just hook on and not look back, but it is just too easy to take out fence posts and slow-moving helpers that way.

4. Gradually approach all gates and be sure of a straight pull through before entering.

Growing and finishing hogs will need a minimum of 8 square feet of sleeping space in a range house, breeding animals at least 20 square feet per head. They will sleep closer together in cold weather, but packing them too close can cause problems, such as prolapse or even smothering with very young shoats.

Double that 20 foot area for a very large boar; a sow and litter may need five times as much space or more. In cold weather, sows and pigs may pack together in these houses, causing pig losses due to crushing from overlay, or smothering.

We like our hog houses to be high enough that we can get to the back without having to drop to our knees. This generally means at

5. When positioning the house, bolster them with pieces of treated timbers or fieldstones to keep the corners raised and free from freezing ground or becoming buried in the mud.

6. In windy areas or with very high fronted houses, it may be necessary to anchor them down after they are positioned. This can be done with simple screw-in anchors and number 9 tie wire wrapped around runners or other structural elements. Righting a flipped house often does more damage than when it was flipped over, so take the time to tie them down if there may be a risk.

7. To keep the shelters even dryer, you can set them atop a mound of at least 4 inches of scraped earth, cull lime, or gravel or screenings. This can then be bladed up when the house is next moved.

8. Once in place, dig a shallow trench around them to carry any runoff water away from the beds.

9. Inspect the house often and keep the runners clear from any buildup of mud.

As long as it isn't overused or abused, the range house will fill a great many housing needs. We've used a small floored house fronted with an oak slatted pen of the same dimensions to wean pigs in cold or rainy weather. A similar unit has proven its worth in growing out young boars for customers that wanted boars with some sort of "confinement" legs.

least a 4 foot high back wall. This is also a must with bigger hogs, such as sows and older boars.

Over time, range house design has evolved, and one design has emerged that does a better job of containing body heat, keeping sleeping beds dry, and being a bit more durable in handling. It is still a three-sided structure on two runners, but the open end is now one of the narrow ends.

Initial construction costs will be greater, as you will be enclosing both of the long sides, and the roof will have to be longer and have a more gradual slope. They also tend to be a bit more difficult to work with, and all the corners will not be as easily accessible for quick inspections.

The trade-off, however, is an overall better structure. The deeper, more enclosed area keeps sleeping beds dryer, and the design does a better job of containing the heat generated by the hogs. The structures are also more stable under tow, and many believe them easier to turn and position. Our first experience with this design involved some 6 x 10 and 6 x 12 houses made from scrap lumber, and they worked quite well. They also seemed to provide deeper shade in warm weather. However, without an insulated roof, a low-ceiling metal roof house in a direct sun field can actually become a hog roaster of sorts.

Range houses more often fail from poor handling than for any other reason, and there are decades-old houses in the country still in position to attest to this. I once attended a farm sale where over 60 range houses were arrayed for sale. There were many sizes and designs. A few were large enough for a man to walk about in, and some were totally enclosed, save for a single 4 to 6 foot wide door. Such houses can be as individual and unique as your tastes and pocketbook will dictate. They are a proven type of field housing that works for nearly any use around hogs.

Range Farrowing Units

For farrowing on range, the most common housing options are typically variations on either the single sow and litter unit or the modular, pull-together farrowing house. With a few modifications or additions, they can be used in every season of the year. We've worked with both types with equal degrees of success.

Both have their fans and their critics, and they both cost about the same. The one-sow units may have a bit more flexibility for handling and portability, but the pull-togethers are perhaps a bit easier to seal up for winter farrowing.

The one-sow units can be as simple as a three-sided and floorless model, suitable for warm-weather farrowing, or as complex as Isolettes, with baffled entryways and insulated walls. These days they can run from under $100 to well over $1000 per unit, and with good care, they'll last for 8 to 10 years. There are some Smidley-type houses (see below) around the country that are crowding 20 years in age and still giving dependable service. A friend of mine from high school once developed a rather profitable business of rehabbing these units for resale. He even set up a hog house lot, much like a used car business, down at the county seat.

The more inexpensive one-sow models look either like cracker boxes turned on their sides or mini-Quonset huts made of lightweight metal. The 5 x 7 foot Pike County Hut (it takes its name from Pike County, Illinois, where many hogs are still produced out-of-doors) has one open end, is floorless, has an insulated roof, is about 3 feet high, and has a sloping back wall. Three or four can be hauled in the back of a pickup, and they can easily be seeded across a pasture for summer farrowing. A really big sow can move one around on her own, sort of in the fashion of a turtle's shell.

These floorless models are inexpensive but are limited to warm-weather use only. They provide little in the way of little pig protection, other than simple guard rails along each side. They must be protected against runoff entering the houses, either by trenching around them or positioning them on a low base of cull lime or the like.

Most new floorless huts run $90 to $150 per unit, and about a third of that for used homes. Most are 5 feet wide by 7 feet long, although I have seen them down to 54 inches wide, and a few up to 7½ or 8 feet long. Once set, they should then be staked or tied down, as they can be flipped over by strong gusts that normally accompany summer thunderstorms.

These are the least durable choice of farrowing units—you really can't call yourself a veteran producer until you have a couple of squashed huts sitting around the farm. We have had a sow or two over the years that has wandered off to farrow somewhere in the great wide open. It's times like this, however, that you really appreciate one of these lightweight units, which can be dropped over the nomadic mother and her new family. These inexpensive huts are often used by beginners and are more than adequate for the true seasonal producer.

The real workhorse housing option for the range producer is the one-sow house with a floor. These are indeed houses, not huts. Typically, they measure 6 feet wide by 8 feet long, 40 to 48 inches high, and have a sliding or lift-up roof.

They are made in a variety of designs and often are called Smidley-types, after one of the more common, bright orange models made by Marting Manufacturers. Those units have moderately sloping inward side walls, a lift-type insulated roof, and are made of tongue-and-groove construction. A commercially built house will run $400 to $600, and some are now even made from injection-molded synthetic materials. Many good used houses can now be bought for $100 or less.

For the record, I have no interest in any hog house manufacturing firm and prefer not to recommend any particular model. I do like to see shelters with three or even four 2 inch wide runners and $1^1/_2$ to 2 inch thick flooring. Plywood in the floor can become quite slick and can cause injuries to both sows and young pigs. Offset doors on each end are a real plus for pig access. The back 2 feet of the house should be enclosable, to form a pig bunk. There, the little pigs can be given supplemental heat safely away from the sow.

We once built a set of these houses that were 10 feet long by 5 feet wide to make use of some 10 foot runners we bought at a good price. They worked quite well but were a bit more difficult to move about and position. We have worked with several factory-made houses over the years, and a couple of firms were even established nearby to build portable hog housing. Most, however, needed a bit of modification for our "tastes," and we've found that the ones we built were probably the most versatile and the least costly, too.

Sliding roofs and straight sides are perhaps the best choice of design elements for the house builder. Be sure to heavily reinforce the framing at all four corners and around the doorways. Doorways are sites of special wear and stress when sows are being contained. Floors, sides, and ends need to be free of harsh edges and sealed tight to prevent drafts from seams in the walls and below.

We have used a lot of native oak for floors and runners, and exterior plywood for sides and ends.

Insulation in the roof is a must. However, the temptation is always to make these houses tighter than they really need to be. A sow and her litter will produce 6000 BTUs of heat per hour, and this is "wet" heat. Installing vent holes in the roof or above doorways can do a lot to keep the houses from becoming little saunas on the prairie. A simple device for regulating these vents can be made by screwing an old tin can lid over them such that they can then be pivoted to adjust the size of the opening.

The farrowing houses should be moved once the litter is weaned. Any wastes can be bladed away or gathered up and spread as fertilizer, and the houses should be given a thorough cleaning and scraping. It was once a common practice to disinfect farrowing quarters regularly after every use, but one school of thought now holds that it is better to get the sows in place there early and let them start building up natural resistance to any potentially harmful organisms in the environment.

When positioning these houses, be mindful of problems with potential runoff and/or mud buildup. These units will be home and hearth for the sow for as much as 2 weeks before farrowing, and then it's home for the litter, too, at least until weaning, 5 to 8 weeks after birth. Some range producers will even leave the weanlings there and remove the sows to reduce the stress inherent in that process.

This will be the hub of the little pigs' world for quite some time, and this must be borne in mind as you ponder design and construction of the shelters. For example, excessively rough surfaces and exposed construction elements may cause navel infections and potential umbilical ruptures as the little guys teeter and scoot around. Chilling drafts can also strike them through the smallest of cracks and openings. The builder must always view and appraise his or her efforts at hog level, not from shoulder height.

Doorways will need some type of low baffle to contain the pigs until their strength and reflexes can match their youthful, wandering ways. We have had little pigs get out the door too soon and die in the mud, get chilled, become lost, or try to graft themselves to other litters. Until about 2 weeks of age, little pigs need to stay at home. We have used 1 x 4 lumber turned on edge to keep the little guys in the house, with fairly good results. We've had a few, however, that wouldn't have been daunted by 12 inch high boards. To prevent damage to the sow's udder, pad the edge of the board with a piece of split rubber or plastic hose. A neat idea I have seen is to fit the doors with individual roller units salvaged from old warehouses or sawmills. They turn when the piglets try to climb over them and they are tumbled gently back into the house.

Heating Options for Winter Farrowing

Depending on the location, these houses will be adequate to use for free-range farrowing for 8 to 10 months out of the year. To use them for winter farrowing, there are numerous steps and modifications that have to be applied. I have seen truck tarps pulled down over them to help better contain body heat. Likewise, I have seen straw bales packed tightly around the end of the house that holds the pig bunk. One anxious producer hereabout even buried his two houses in deep mounds of naturally insulating snow.

A lot of producers will pull them inside a clear-span building to provide further wind protection and containment. A common practice in our area is to pull them into a central location where they can open out onto some sort of floored pen, and where electric service can be provided. A number of supplemental heating devices are available that can be suspended in the pig bunk area.

Most common, of course, are electric heat bulbs in simple metal reflectors. They are available in 125 and 250 watts. The red-tinted varieties seem to do a better job of drawing the pigs away from the sow and to the security of the pig bunk. In extremely cold weather, we have used as many as three bulbs, with one hung over the farrowing sow and two in the pig bunk. Needless to say, we were giving this sow near-constant supervision.

Heat lamps should never be suspended by their cords or tied into position with baling twine. Suspend them using either heavy-gauge wire or lightweight chain, well away from the reach of the sow and far above the reach of the little pigs. They seem to get a kick out of popping the bulbs by touching them with their little wet noses.

The bulbs should be suspended a minimum of 24 inches above bedding and away from any wooden surfaces. The bulbs and reflectors should be kept free of dust. Inspect them often when in use, and take them down and put them in storage when not needed. There should never be more than seven heat bulbs plugged into a single circuit. I've seen young pigs that were badly burned from heat lamps that had slipped or had been neglected.

There are some portable gas-burning heaters also available on the market. One type feeds gas into a ceramic element that is flameless but provides a good deal of heat. The gas can be supplied from a small tank attached at the back of the house, beyond the reach of the sow. This is a viable heating choice for houses left on range. It is probably the costliest among the supplemental heating options, but it will work where electrical service is not possible.

We used individual houses two different times over our 35 years with hogs. We started with stalls for farrowing built into a centralized barn on our first farm. From there, we went on to eight crate pull-togethers. When we moved to our current site, we went again to one-sow houses.

Over the years, I have gotten fairly good at skidding these houses on and off of pickups and have towed them many a mile behind our old H and Moline tractors. When it was necessary to fine-tune their placement, I have even rolled them on short lengths of 3 inch pipe and nudged them along with an old railroad pry bar.

Variants on this type of unit include A-frame houses, modified A's, and double units that capitalize on the increased body heat output of two sows and litters. Plans for many of these types of housing can be found in old ag texts. One guide we found to be especially useful, with very detailed building plans, was the old 4-H woodworking manual that came out in the '50s and '60s. If you find one of those old books, hold onto it tightly, as it stores a wealth of good building ideas.

Once you have put them together, you will find that one-sow range houses qualify for rapid depreciation (for the tax rolls), that they can change farms with you—you certainly can't dig up a swine lagoon and take it along when you move from one farm to another—and that they have substantial resale value. These small houses can also be used with any number of other livestock and poultry species.

The Pull-Together Farrowing House

One farrowing house option is the modular type, which can resemble a small building, when set in use. It is essentially a building on runners that has been built in two halves, much like a double-wide mobile home. The halves are then pulled together; hence the name, the "pull-together" farrow house.

Theses are essentially 7 foot wide units that stand in lengths anywhere from 10 to 25 feet. They are built with shed-type roofs that have 3 to 4 foot overhangs to form a covered center passage way or "alley" when the two halves are pulled together. This center alley is generally the primary access point and service area. Each half is built on two 4 x 6 to 8 x 8 foot runners that extend the entire length of the halves.

These are the basics of the pull-together's design. Beyond this, the sky really is the limit. They can be made as simple or as elaborate as you have a mind to create. The most basic unit out there is composed of two 7 foot halves that are floorless. With a half-dozen short gates, they can be cordoned off to form four 7 x 8 farrowing stalls. One corner can be gated off to form a triangular pig bunk, initially allowing the sows to be turned out for feed and water, and then full-time shortly after farrowing.

At the other end of the spectrum are floored halves containing four to five 5 x 7 foot farrowing crates in each half. For many years, we worked with an eight-crate house that was built for us by members of a nearby old order Amish community. The runners, floor, and frame were all constructed of native white oak, and the sides and roof were made of corrugated tin. We wired it ourselves for electric heat lamps

Farrowing Crates

Farrowing crates are a hot-button issue for some producers, and to be sure, they do restrict sow movement to a fair degree. Their original intent was to save pigs from overlay, not as a means to pack as many sows as possible into a building.

We've used crates made with 2 inch oak planking that provided the sow with a lying-in area 2 feet by 7 feet and an 18 inch wide pig bunk that ran the length of the crate down each side. We experimented with different widths, and with some older sows did get to 26 inches or a bit more in width. With gilts and young sows, this additional width would sometimes tempt them to turn or climb up on the side rails.

This is a new generation angled crate that gives the sow greater space and ease of movement — she can even turn. It is even more comfortable with the addition of a good bedding of straw.

In very severe weather, we have kept sows and litters in the crates for up to 2 weeks after farrowing. The sows were let out for at least an hour each morning and evening, away from the pigs to stretch their legs, drink freely, and do a little basic lounging.

We would move the sows into the crates 5 to 7 days ahead of farrowing, to ease them into the routine of life there. Our goal was always to get the litter out of the crate within 7 to 10 days after farrowing, and into much larger nursery pens on range or in drylots.

The traditional farrowing stall was 5 by 10 feet, with the rear 2 feet of the stall gated off for a pig bunk. We built a set of these into an old horse barn on the homestead, and they fit quite well into barns of that era. We used hinged-drop sides made of plywood to open up or enclose that bay of the barn as weather and season dictated. The sows had a slightly greater degree of free movement, but housekeeping chores and pig losses were greater as well.

The farrowing crate is very much an individual call, but like most other things, is neither all good nor all bad. It can be abused—I certainly wouldn't want to see a sow share one with a litter of near weaning-age pigs, nor a pig confined to a crate 24 hours a day.

For ease of use and animal well-being, the crates should be stoutly made. All surfaces should be free of rough edges and any raised hardware. Inspect them for damage each time the sows are turned out. Our crates were on an oak floor, and we would use a little straw bedding for added sow comfort and to control dampness.

and set it adjacent to our sow pens for year-round farrowing here in northeast Missouri.

In this house, the sows entered and exited the crates through the center alley. A large door was hinged to each half, and when the halves were joined, the doors could then be closed to seal the building. Twice a day, we turned the sows out to eat and drink in a small lot that we built adjacent to the house. Supplemental feed could then be provided to each sow in her own crate, depending upon her health and litter size. Crates could be cleaned and the pigs tended while the sows were out. Given a regular schedule, most sows become good at relieving themselves when turned out into a lot.

The ultimate in pull-togethers was a series of 10 and 12 crate houses I saw a few years ago. Along the outside of each half was a slat-floored platform holding a 5 x 7 pen for each crate in the house. There was a swinging door in the outside wall for each crate, and the sows and pigs could come out on these sun porches fairly soon after farrowing.

Pull-togethers are perhaps the least mobile of portable housing options, but relocating them isn't as difficult as, say, brain surgery or juggling chainsaws either. By the numbers, it goes something like this:

1. Use a tractor that is adequate for the task. A 50 hp tractor would move the four-crate halves of our house, but an 80 hp tractor may be needed for some of the bigger units. Also, it's best to wait until the ground is dry or else is frozen solid. Lose traction with one of these, and both you and the building are going to be there awhile. What's more, you can't exactly tell pregnant sows to hold off birthing because you've got the farrowing house stuck in the mud down by the far pasture gate.

2. Make sure that the house is completely free of the ground before starting the pull. The same is true of water and electric lines. Looking back to see streams of water and sparks flying is simply not good form.

3. Move the halves using a slow, steady pull. No jerking, even at the start, and no sharp turns.

4. Carefully plan out your route of travel, avoid tight situations, and position the houses with a thought about their next move.

5. When the house is set, position heavy timbers between the runners at the ends to block runoff and to keep chilling drafts from entering beneath the floor. Build berms or trenches as necessary around the house to direct runoff away from or around the house.

6. Moving a pull-together isn't something I would want to do every day, and it certainly is a job best handled with a couple of helpers. If moved properly, they can be shifted around without any damage, so there's no truth to the old tale that a pull-together can be moved only so many number of times and then it's shot. Do it wrong and you can tear an end out of any brand new house.

One modification we made to our house was to construct a platform floor in the center passage, using 2 inch oak. It prevented any mud from forming there and made it a better area from which to work and tend the pigs. We sometimes extended the pig bunks onto this floor to facilitate creep (self) feeding for the little pigs.

The roof was insulated with 4 x 8 sheets of black insulation board. It was also used in the sidewalls, where it was shielded from the hogs with 1 x 2 inch wire mesh. The house was given a cleaning and scraping after each farrowing, and we threw spent bedding to the next set of sows to go into the house. The contact with spent bedding allowed them to develop a natural immunity to any potential organisms in the house and pass this resistance on to their newly born pigs through their colostrum.

A new pull-together will set you back $200 to $300 for each crate or stall it contains. This is for a basic unit. Expect to pay a third to a half of the cost of a new house for a used unit. Transportation can be a substantial added cost with these units, however. Figure at least a dollar a loaded mile if you have to have one hauled, and it can go up substantially from there.

Hog housing is a tool, one that is more of a means to an end. And that end is profitable swine production. The only people who make money from the hog housing are the builders. Your task is to make money using it.

Right now, there are many farms across the Midwest that are actually crippled in value because of the hog facilities that were built and remain there. This is because the infrastructure they have set up depends on suspect waste-handling systems that dictate and thus limit how a farmstead can be stocked and operated.

Animal traffic through confinement facilities must be pushed at an accelerated pace to keep the banker happy and to satisfy overhead costs. The numbers have to flow, regardless of input costs or selling price.

Of course, it matters not to a hog whether it was born in a $20 hut or a quarter million dollar building. In the end, what matters is that there is housing that's adequate to the animal's needs but that doesn't detract from the bottom line.

Range production is low-input agriculture, and housing should be considered effective if it maintains the animals comfortably and securely.

The Sick Pen

One special housing provision that should be in place for every farm with livestock is a sick pen. Simply, this is a tightly-made house that's set well away from other hogs, where sick or injured animals can be isolated for medical treatment and their own safety.

The best sick pen we ever had was a good 6 x 8 house fronting a 12 x 12 pen atop an oak platform. It was set about 100 yards away from any other hogs on the farm and rested under the shade of a big oak tree. We moved animals to this pen as soon as anything irregular was detected. Plenty of water was offered in an easy-access trough, and the house was kept well bedded. The pen was enclosed with 4 foot high plank gates set up with steel posts on 4 foot centers. Those hogs weren't going anywhere and could be safely restrained if crowded against a side of the pen.

The pen wasn't used all that often, but when it was, we really came to appreciate it, out vet included. It was the last pen tended during morning and evening chores, and boots were cleaned thoroughly just to be sure nothing was tracked from pen to pen.

Feeding & Watering Equipment

I have known my dad to spend the better part of a week building a pig creep feeder, a walk-in model, and while it was about as plain as homemade soap, it certainly stood up against even our most determined sows. And that is the key to swine feeding and watering equipment—it has to be durable.

I have seen feeders with the bin part actually made from concrete. I have also seen them emerge chipped and battered after just a couple of years on a high-volume feeding floor.

Our first watering appliances were troughs made from old hot water heater tanks cut in half lengthwise. Legs and cross-members made from steel straps were welded on to give them stability and to keep the hogs from wallowing in them.

Our feeders were V-shaped troughs made from 2 inch thick planking. As long as we have had hogs, we have never been without at least a few troughs, some of them metal 2-footers that are near-priceless for use with nursing sows.

Hogs will often drink out of cow tracks and eat off the ground, and a lot of sows on range are still fed by bucketing out feedstuffs along high spots in the sow lot. Some of the best exercise I ever got was by running ahead of a pen of sows with a 5 gallon bucket of feed in each hand. One or two sows can be fed in simple rubberized tubs or pans that can stand up to the worst kind of abuse. They might tear, but they'll never crimp or dent.

Sow feeding will always be the greatest challenge, and especially so in the rainy and muddy seasons. I've seen sows spend hours slurping their way through an ugly broth of mud and snow and shelled corn.

An illustration of the deep-bedding system now used in cold-weather feeding in hoop houses. As the straw breaks down and packs, it will even generate some added heat. Fresh bedding can be added over the top of the bedding sack as needed. When the hogs are moved, the bedding can be scooped us and spread on crop ground.

They probably didn't have anything better to do with their time, but it certainly is wasteful.

In some sow lots, I have seen feed spread on concrete pads. A few have even had trough spaces pressed into the concrete.

Ideally, feeding stalls will allow each sow to be fed individually and to get their fill. These are generally sections of four 2 x 6 stalls built atop an 8 x 8 runnered platform. They can be hauled around and positioned along fencelines where the sows can enter from pen side, and the feed boxes at the front of the crates can be filled from outside the pen or lot. As the sows enter the stalls, a gate should be closed behind them to prevent fast eaters or bossy sows from finishing and then pushing other sows out of place.

We've always built our stalls out of oak, in a four-sow configuration. The partitioning gates should be at least 4 feet high and made with full-cut lumber at least 1 inch thick. Over the top, we nail a simple flat roof of sheet tin to keep the sows and the feed dry in case of snow or rain. Feeding-stall segments are then matched to pen populations.

Sow groups can be fed in shifts to maximize feeding-stall utility. They can be positioned in a gated common area where one set of sows at a time can be allowed to eat. There must be one stall for each sow in the group, however.

Many producers will build one extra-large stall to accommodate the boar if the stalls are being used with breeding groups. We simply closed the sow stalls with a swinging gate and then fed the boar in the open area.

While never easy, the task of fitting new females into an established group can be simplified somewhat with the feeding stalls. The younger and/or thinner females can be given the extra feed they need in the

The Boss Sow

Hogs in a group very quickly establish a clearly defined pecking order. In the breeding herd, the top position generally belongs to one of the older and larger females. She will continue to direct things, even when the herd boar is present in the pen.

She uses her bulk and experience to get her own way and yet she will also be somewhat protective of the group. The smart producer can sometimes use her position in the herd to his or her advantage. If you can get the boss or dominant sow to do something, the rest will generally follow in good order. Remove her and you will create a small civil war until a new order becomes established.

The quality of construction and the materials you use should be able to withstand any kind of punishment from this sow. Since she is usually the biggest female, she will be the one applying the greatest push against walls and gates. And when she rises from a bed or passes through a doorway, the rest will soon follow, generally en masse. If you build your doorways, gates, and corners to be more than a match for the wiliest of boss sows, you will have few problems with containment and care.

If you don't, well . . . one morning we went out to find a group of sows playing soccer with an empty tank waterer. We also had a couple sows that Dad swore could hear the instant an electric fence charger went offline. On the other hand, we had a couple sows that were our equals in taking care of the sow herd.

You will always have one boss sow, so learn to know and respect her. Hers is a role, an important one, that has been in existence as long as there has been a hogs on this Earth.

stalls, preventing the more aggressive females from taking feed from the others and getting too big. This is also a good place for administering certain medical treatments. In the stalls, you can make sure each and every female is treated correctly for parasites, for example. They will add a few minutes to the chores, but they give an element of control over the breeding herd on pasture that cannot be achieved any other way.

A big sow will easily drink 5 gallons of water in a day, and often a good bit more during really hot weather. I have measured this myself time and time again in trips with 5 gallon buckets carried up from our spring to the sow pens. The way that we replenish water has to be one of the simplest, most effective watering options: a steel trough that's been slid under the charged fence and into the pen about 18 to 24 inches. The larger part of the trough remains outside of the pen, and we fill from there once the trough has been firmly staked into position. The hogs can drink from the short side of the trough, but the wire keeps them from lying in it or otherwise wasting the water. The troughs are also easy to move if a mud problem develops.

On the retail front, there are all manner of hog waterers out there, and all will work, though each have their own little quirks to deal with. One of our favorites has been the 80 gallon round tank with a water dispenser, or "fountain," on each side.

These we would set into a fenceline at the end of two adjoining pens. Each fountain is adequate to serve 10 to 25 animals. They fill from the top and generally have a conical lid that will blow all over creation unless you take the time to fit it back tightly. Many of these have a second bottom into which kerosene burners can be fitted for use in cold weather. Just be sure that you position the tanks with the access doors turned out, and always set them atop a platform that extends at least 2 to 4 feet beyond the fountains.

With any number of animals, these will need to be topped up daily. Should mud begin to form around them, either move them or fill in the area with stone. We have used a lot of creek gravel in hog lots over the years. However, these days you should check with your state's Department of Natural Resources about current regulations for collecting creek gravel.

Another common choice for waterers is the oblong tank with fountains recessed into the tank's sides. These types will vary in capacity, from 100 to several hundred gallons. Fountain configurations will vary from a single outlet centered on one side of the tank to two offset fountains, one on each side, to two fountains per side. The larger the tank, the more fountains that can be fitted into it, up to a maximum of four. More than that, and the sidewalls may become too weak. These tanks are easily moved about and can also be fitted into a fenceline, although they may need a bar or two over them to discourage jumping.

One of the greatest challenges to range producers is keeping that water liquid in the wintertime. There are electronic waterers that are fitted into concrete pads and hiss and whir and feed heated water into a steel drinking cup. They're the high-dollar option and seem to breakdown only in the deepest cold, when they are needed the most.

Watering tanks can be painted a flat black to draw as much heat as possible from the faint winter Sun. It is solar heating of the most passive sort, but every little bit helps. Also, to this end I have seen some black membrane bladders containing a bit of chlorine bleach positioned atop watering tanks. Also available are a number of immersion and floating-type electric tank heaters.

These come with fair lengths of armored cable but must still be carefully positioned to be safely beyond the reach of the most curious hogs.

Hogs are always going to be harder on watering equipment than any other large animal species. The surface-based, gravity-fed systems that we now see supplying water to cattle on rapidly rotated pastures would be a disaster waiting to happen in a hog lot.

Watering isn't exactly the Achilles' heel of the range or pasture system, but it gets close. Some range producers will need a tank and a truck or trailer to haul water, some may use gravity and buried piping, and still others will have a few pressurized hydrants and hoses that they rely upon.

For us, it was long a part of the winter routine to fill tanks using a couple hundred feet of heavy garden hose and then walk it out, lifting and draining it as we walked. It was cold and wet work, and still we often didn't get all of the water drained. That left us with a frozen spot to stomp on or try somehow to get open the next time we used it.

For growing and finishing hogs, you should provide a fountain or at least one foot of trough space for every five head or so. Hogs have a tendency to want to do things as a group, so if trough watering just a couple of times a day, provide plenty of space and be sure everyone gets enough to drink. Self-waterers are a far better choice for growing hogs, however.

Water is the single most important provision and must be supplied to the hogs in a clean, dependable manner. It is critical to all phases of swine growth and development and should be job one in plotting you outdoor swine operation.

Feeding Equipment

At one time, all hogs were fed once or twice a day by hand. The feed was poured onto the ground or into troughs. It is still the way a great many brood sows are fed.

I recall one year when things were a bit tight, Dad caught me really stringing out the feed in front of our old sows. He reminded me that by making them work that hard and walk that far, I was really causing them to need more feed.

We often hear the story of the old farmer who, before the Big War, was shown his first self-feeder. Upon seeing it, he scoffed and dismissed it as ridiculous, since with it they were sure to spend all of their time eating. Now, I'm not sure what other pursuits he had in mind for his hogs, but when mine are eating, they are just moving closer to market.

Little pigs learn to eat dry food by watching their mothers, and their first solid food is generally bits snatched from their mother's feed. I've seen little guys try to root their mother's away by the time they are a couple of weeks old. A variety of starting feeds are available for young pigs, and these are discussed in greater detail below. Starter feeds should be offered to the pigs and the pigs only. Sows and raccoons both love pig starter, but it is not a cost-effective ration for either creature.

The first feeder for nursing pigs on range will be a walk-in or gated unit that will admit the young pigs but keep the sows away from the very palatable—and rather costly—feed. Little pigs will eat only quite modest amounts of this, generally consuming no more than 50 pounds in reaching a live weight of 40 pounds. It is nutrient dense,

A creep feeder that allows young pigs access to the feedstuffs, but not the sows. Note that the feeder and waterer in the foreground are positioned on concrete pads to prevent problems with mud buildup.

These feeders are immediately adjacent to the fence line for easy fill-up and access.

manufactured in forms easily consumed and digested by young pigs, may be highly flavored and aromatic, and should be changed often to keep it fresh and appealing.

Young boars receive a full-fed grower ration from a self-feeder. They are also receiving the nutritional advantages of being on good pasture. These developing males will benefit from the ever-varying and stimulating range environment. Pictured are young Duroc boors of good type and breed character.

Most sows see a creep feeder as a challenge with a flavorsome reward inside. Few items on a farm with hogs are more banged and battered than the creep feeder, and although they're built to be sow resistant, time teaches us quickly that there is no such thing as "sow-proof."

On range, where they will be less frequently attended, it is best to gate off any pig feeders. The young pigs can be allowed to enter and exit the pen through an 8 inch high opening at the bottom of one of the gates. Set those gates very securely, as the sows are sure to give them a workout.

These types of walk-in feeders can even be used by the pigs for a time following weaning. They will accommodate pigs up to 40 pounds in weight, which can then be moved to a self-feeder that will accommodate them to market weight. In the first few days on this feeder, be sure to tie a few of the lids open until the pigs get the hang of how the feeders work.

Self-feeders are manufactured in a variety of styles, the most common being the round type, with capacities of 40 to 80 bushels of feedstuffs. These are basically a round bin atop a stainless steel or cast-metal cone that funnels the feed into a heavy metal base. Surrounding

Buying Used Equipment

The farmer's version of a busman's holiday is to go to a farm auction and hunt for bargains. I like nothing better than to attend a sale and sort through the junk wagons and shake the corners of used buildings.

It makes good economic sense to buy used if you're buying with a plan. The first rule of thumb when buying used is never to pay more than half of the new price unless it is of very exceptional quality. In figuring the costs of acquisition, don't forget to factor in the transportation costs to get your purchases back home. In most rural communities, there is at least one owner of a flatbed truck who has the skills to haul anything from steam engines to hog houses.

In buying used, give yourself plenty of time to inspect the items and thoroughly evaluate them. For example, if they are being offered on a farm auction, go the day before and really look them over, without any of the normal sale-day disturbances. You may even get to glimpse items being moved about and have more time to visit with the seller. You don't want anything that has been exposed to disease problems or that is being sold where it's stuck down in the muck of an old hog lot.

At the moment, used hog equipment is priced down sharply in many areas, but some of it was also long past replacement and badly abused when hog prices tumbled a few years back. Good used equipment should still have many years of productive life ahead of it. I have crawled into neighbor's junk piles to retrieve replacement feeder lids or waterer tops, but dollars are too hard to come by to throw away on equipment that is better left to the junkman.

The hard and fast rules of buying used are to thoroughly check over a prospective purchase, pay for the useful life left in it—and no more—and to remember to include the costs to get it home. All used equipment should be presented for sale thoroughly cleaned and ready to load out and move on.

When we buy something used, we will give it a second cleaning on the spot. Use a scraper and stiff broom to clear away any remaining dried mud or old feed. Many buyers also pack along a small sprayer with a disinfecting solution to spray down their new acquisitions. Others will stop by a local garage on the way home and have their purchases steam cleaned or pressure washed as a further sanitizing step.

With used items on the farm, I like to borrow a step from the old-timers. We drop the items off well away from any hogs, open them up as much as possible to the naturally sanitizing rays of the Sun, and then leave them there for at least 30 days. Sunlight and a good airing was my grandmother's cure for just about everything, and it works equally well for hog equipment.

Hog equipment and housing that has stood unused for any length of time begins to go down for reasons I'm not sure anyone can fully explain. It is a phenomenon comparable to what happens in houses that are left empty. Just beware of those items that have been pushed deep into a fencerow and left there for a long time; they don't bounce back very well.

Do your best to buy the best you can.

the base are a dozen or so lidded feeding outlets. Each one of these lids should be able to accommodate three to five growing hogs. They have slide-back or lift-off lids, filled through the top.

Feeders are always high-traffic areas, so they should always be positioned on square platforms or concrete pads at least 12 feet on a side. They can be a bit top-heavy when in use and may topple if the platform shifts in the mud. Move them often if they are to be positioned completely inside a pen. For the wear and tear they receive, they are quite durable and can be almost entirely rebuilt, if need be. Extra rings can even be bought to increase their feed-holding capacity. Rubber feeder lids have proven to be more durable and are much quieter in use.

The other common type of self-feeder has a box-type bin with feeder lids down each long side. These are true fenceline feeders that can serve two abutting pens that contain hogs of the same age and size. These have been made with both wooden and concrete bodies, along with galvanized metal models. Due to their narrow base, they are more difficult to move about, and just like the round models should be mounted on a platform or concrete pad.

Still to be found around the country are the classic step-in feeders, which will accommodate growing hogs to butcher weight. They might have been the most stable of all feeder designs, and with their lower lift-tops were the easiest to fill. Had they been made with heavier-gauge metal, I believe they would have become the standard for self-feeders.

Feeders and waterers have a twofold task: first, to put needed feedstuffs where the animals can get at them, and second, to keep those feedstuffs clean and fresh. You can get these in plain or fancy types, but these are two things that they must do. With good management and care, they can last for years.

Most feeding and watering equipment is now made for use inside confinement units. They have small bases that fit into the tightly packed pens and are refilled frequently from augers, drop tubes, or pressurized piping. It is replaced often and is usable nowhere else but indoors.

The small-farm range producer will, for at least a time, have to be working with a fair amount of second-hand and refurbished equipment. The good thing is, a lot of this sort of equipment remains on the nation's farms. We made our own start largely with used equipment, since it is the tried and proven method of paring inputs.

Handling and Sorting Equipment

A few years ago, the vet and I were crowding a Duroc sow to the far corner of a pen to block her there for a needed injection. At the last moment, she turned and launched herself skyward. The good doc and I reached up and luckily deflected her in flight. Realizing that we both had been reaching for her, the vet looked at me and said, "Just what would we have done if we had caught her?"

I don't know what we would have done with 500 pounds of airborne sow, but it does point out the importance of good handling and loading equipment. When I was younger, we could and did crowd sows against a solid barn wall with heavy oak gates, but even then I knew there were better ways. There had to be.

The Catch Crate

From that aforementioned 4-H plans book, we had what's called a "catch crate" built in the local ag shop. Ag and tech classes at high schools are always looking for good teaching projects, and you can often get needed items of equipment built there for little more than the cost of the materials.

The crate was set on runners, had an end and side gates, and a simple locking head gate. It would handle shoats from about 100 pounds up to the largest of sows and boars. It was built with gate-type sides, although solid enclosures tend to make the hogs more comfortable and easier to work with when confined.

The crate could be towed from lot to lot, and in a pinch, we would even use it to move a hog over short distances. A catch crate is useful for restraining hogs for injections, ringing, and other types of care. Older hogs can especially become a bit crate wary, and having enclosed crate sides and run-up will do a lot to allay their fears. Also, do as much as you can while you have them in the chute or crate, to keep from having to run them through too frequently.

Over the years, we have collected an assortment of gates and panel segments of various lengths that serve us well in sorting and handling hogs. You never know when you will have to build a hog corral in the shape of a tetrahedron out in the middle of nowhere. With that done, you then have to back up a loading chute to it and make a pig-tight union between the two. You can never own enough short gates.

Chutes

I knew I was a committed swine raiser when I bought a brand new hog panel just to cut it into short pieces to carry along with our loading chute. I go to auctions and buy short lengths of livestock panels much the way others collect cut glass or art prints. The range producer will benefit greatly from an assortment of lightweight gates that can be moved about quickly to set up holding or sorting pens all over the farm. Set the pens a few days before they will be needed and offer a bit of feed there to get the hogs comfortable with entering the new pen and spending a bit of time there.

We could fill the remainder of this book with a discourse on what constitutes a good loading chute, and I'm very sure that some day a book on loading chutes will be written. The loading chute is the hog's first and last stop on the farm and is one of the primary sources of injury and agitation.

For starters, here's a simple list of some of the key points of a loading chute design and operation.

1. A lot of chutes these days are built and sold for "all large animal species," which ranks somewhere in between "one size fits all" and "the

check is in the mail." A swine chute should be no wider than 24 inches, as the hog's primary method of escape from a chute is to turn back.

2. Build the chute high enough for you to walk up behind the hogs, always moving them forward by blocking behind them with a "hurdle," which is a flat panel with handles used to move animals along.

3. Floor cleats improve traction, make the animals a bit more secure, and add to the overall safety. The flooring should be at least 2 inches thick. It should be cleaned after every use and inspected often for wear and damage. The floor is the most common point of failure in a loading chute, and an animal breaking through a poorly constructed bottom can suffer severe injuries in the attempt. Let one drop through a chute floor and you will never get him up another one.

4. The ramp up should be as gradual as possible, for both the safety and the comfort of the animal.

5. A loading chute for hogs should have solid sides to prevent any outside distractions.

6. Driving hogs out from a dark building to a chute in the daylight will cause the hogs to balk.

7. Don't rush the hogs, don't lift them by their tails, and be sure to budget extra time for the loading process.

Over the years, we have had stationary chutes, chutes on runners, chutes on wheels, and a few that were little more than a chute body that could be slid out of a truck bed with one end lowered to the ground for loading anywhere the truck could go. A number of methods have been incorporated into chute designs over the years to enable them to match up to truck and trailer beds of different heights. Most have the front of the chute moving up or down on a series of manually adjusted pins and stops.

Walk the chute often to check the floor and look for any broken boards or raised hardware that can injury hogs or cause them to balk in the chute. To ensure a longer life, store it indoors between uses, if at all possible. I firmly believe that no single piece of equipment on a livestock farm is as abused and neglected as the loading chute.

Hog Trailers

Chasing hogs loses its charm and comedic thrill very quickly, so on every farmer's wish list is a hog trailer or pig carrier. They are at once effective swine transport and an easily portable working pen.

The pig carrier is lifted and positioned using a tractor's hydraulic system. It is rear-mounted, raises and lowers with the tractor's hydraulic arms, and has a pen area roughly 6 x 6 feet. These work best with a larger tractor, at least 60 hp. They can be used to transport a couple of sows and litters or a few butcher hogs. Sides need to be 54 inches high or so to discourage jumping when the carrier is raised.

Nowadays, hog trailers are built with the flair and variety of customized cars. Some have the entire bed raise or lower with the tractor's hydraulic system, while for others, the tailgate can be lowered to form a loading ramp. Some have center dividers or side gates to facilitate sorting.

We built a 6 x 12 version on a single axle, with a hinged center gate and rear gate that formed a loading ramp. The bolsters were treated 4 x 4's, and everything else was made of that old and always reliable native white oak. Working the end gate was an exercise in weightlifting, but the rig was handier than a shirt pocket. It pulled easily behind our smaller, two-bottom tractors, and we used it in a great variety of ways around the farm. It was our take-anywhere, do-anything pen. And with all of that oak, it wore like iron. We even had a pig holder made that would attach to one of its side panels, which would enable us to do nearly every pig health chore.

With fencing, housing, and equipment in place, you are ready to begin outdoor production. Together with your plan of production, they form two of the three base points on which to build. The third is the hogs themselves.

You need good equipment to do a good job; it is not an end unto itself, nor is it any real sign of success. Flashy equipment looks good on magazine covers, but may only make some money for the fellow who buys it after the initial buyer goes broke trying to pay for it.

It's important to keep in mind that the hogs are what make it a hog farm, and good hogs are what make it a good hog farm. Good tools, appropriate tools, can help you to do a better job, but a two-bit pig going into a two million dollar barn will still come out of that barn a two-bit pig.

CHAPTER 3

SELECTION & BREEDING

I have an 11-year-old great-nephew, Kendal, who was born to be a salesman. As Dad would say, he could sell you a dead possum in a sack with such conviction that you'd walk 10 miles home believing every step of the way that you had just bought a lively shoat pig.

Fortunately, Kendal also has a very good heart, and we hope he will use his gifts for good things. But there are some less-than-scrupulous people out there, so how do you start a swine operation without walking away from a sale with that "Did I just buy a pig in a poke?" feeling?

Well, to state the obvious, in order to know what to buy, you have to know what you need. The hog that's ideal for outdoor production is not going to be simply plucked from the nearest confinement-influenced swine setup. The genetic package you will need has fallen a bit out of favor of late, and to use the inflective of an earlier day, you're going to have to go a-lookin' for the kind you will need.

While not exactly porcine athletes, these animals are going to have to be able to work for their living. The range hog will have both muscling and growth potential in fair moderation, but they are also going to need sound and true feet and legs, good lung capacity, depth of side, a vigorous nature, and a healthy dollop of body cover.

A hog on range, unlike a beef animal on range, will not be expected to navigate hundreds of acres and thousands of feet of elevation, but it will have to get up and put one foot in front of the other to reach its supper. It must have the will to rise up out of the sleeping bed on a cold and snowy morning and get about the business of being a hog—living, growing, and reproducing.

Sadly, there are hogs out there today that are failing in these most basics of life, and confinement rearing has been a crutch for some rather dubious genetics. There are lines in many breeds that have been engineered for confinement literally for decades. Many breeding lines in from Europe and the Scandinavian countries are especially noted for this. One of the White breeds a few years ago actually spent money boasting of some of its lines that had scores of generations reared in total confinement. Some of the commercial seedstock lines now also carry a heavy dose of Pietrain breeding.

The Pietrain is a double-muscled breed that is about as extreme a type as anything you will find. Like all breeds, it has value, but within the breed are numerous lines with stress and soundness problems. There have also been meat quality issues with some of its lines. This is a second go-round for this breed in my lifetime.

A couple of years back, a market hog with substantial Pietrain breeding was driven out at a local fair. Pietrains will adapt the colors of other Black hogs to which they are bred, but there is no hiding that muscle pattern. The judge did not deride the hog, but he did have it removed from the ring for being far too extreme in type for a market hog class in the Midwest.

Given the numbers in a litter and the short time between generational rotations, you can breed a hog to look anyway you want it to in fairly short order. This explains just how quickly Red and White Pot-bellied pigs were able to appear on the scene when that fad was hot. This ability to affect change in a line can be a good thing but is easily abused by those who chase fads for their own gain. One old hand once confided to me his belief that changes in type came every time the big boys needed something new to sell.

About 25 years ago, the swine industry was swept by a breeding trend toward flatter muscled and slower growing "later maturing" animals. It would take more time, it was reasoned, to grow the necessary frames on the big butchers that were and still are envisioned as mini-beeves. The result, however, was a lot of cat-hammed "meatless wonders" that were hard to breed. It was the fad that opened the gates to the factory-type seedstock companies.

They used that oldest of breeding strengths, heterosis—hybrid vigor—to create breeding animals that would live well and breed.

Those, my friends, are the most important of all economic traits. They marketed them rather like new pickups, with all kinds of features, and even used factory showrooms. There may have been a lack of uniformity in their offspring, but by golly there were offspring.

Soon after that, we began to see a barrage of facts and figures: performance data, herd indexes, carcass scans, and enough charting and graphing to wow the folks at NASA. Numbers can be boosted in any number of ways, their meaning and value lost in a sea of competing and questionable data.

At one time, it was held that the show ring was the best arbiter of swine type. There, producers would submit their idea of what a good animal should look like, driving one out in a head-to-head competition between peers and for everyone in agriculture to see. It is almost democratic in its appeal, and it is a practice with deep roots going back to the age of medieval trading fairs.

Hog shows have gone on to produce a fair number of barrow-show jockeys who could primp and preen even a so-so hog into a blue-ribbon winner. One of my favorite pastimes is to find a seat behind some old-timers sitting ringside at a hog show and listen to their running commentary on hog exhibitors and hog-show judges.

Over time, many have come to question the value of the show ring, in light of the more industrialized nature of confinement production. Also, a growing number of "show" animals are to real hogs what an Indy car is to your old pickup. Still, I believe the show ring retains a lot of value.

It is a good training ground for young people, it's an important social hub for the community of swine raisers, and it still parades the latest in types before the unrelenting gaze of hard-eyed veterans. And there, hopefully, the youth will walk about the show ring listening while the judge talks about the animals. The places where you see the most hogs—fairs, type conferences, sale barns—are not the best places to buy hogs, however.

Trends in type seem to move through the swine industry faster than fads among teenaged girls. There was an average of at least one major change in swine type every 5 years in the 20th century. An estimated eight swine breeds have become extinct in this country in the last 110 years. One, the Ohio Improved Chester, was lost as recently as the mid-1960s

A couple years back, I was part of an informal panel on livestock production and the family farm. Around those tables were a noted magazine publisher, a legendary Hampshire swine breeder, a noted expert on range poultry production, and me, with 30 plus years in the hog business. One of the first questions just happened to be about the selection of foundation animals for an outdoors operation. We all took the bit in our teeth and ran with that one.

The first point we all agreed on was the importance of purebred livestock to the independent producer. Too often lately, breeding stock selections are made for producers by pork contractors and support industries. If you are producing hogs under contract, one of the conditions of that contract may be that you have to begin with the purchase of a herd of ABC line gilts from a seedstock company allied with the contracting firm. To maintain any degree of predictable heterosis, you must then breed those ABC gilts to boars from their XYZ line of terminal males. Their offspring must then be bred to a second strain of MNO boars, and you must start again with another set of ABC gilts bought from off farm. Where is the independence in any of this?

If preserved and propagated wisely, purebred livestock is the genetic resource that will remain forever with the family farmer. For 30 plus years, we have worked with purebred hogs, and they always gave us a sense of having a full say in our own destiny. It was and remains America's purebred gene pool that the seedstock companies draw upon to formulate their composites and breeding lines.

The White pure breeds are termed the "mother breeds." Their historical strengths are mothering, milking, and temperament. Their reputation is built on large litters and exceptional pig survival. The colored breeds are generally termed the "performance" or "meat" breeds. They are hardy and faster growing, coupled with a larger and more defined muscle mass. All of this being said, however, you must always keep in mind that most basic of all truths about swine breeds: the differences between individuals within a breed is always greater than the differences between breeds.

With the single exception of litter size at birth and at weaning, our Durocs and Chester Whites were virtual equals on all other fronts. Our first goal was litters of eight to 10 uniform pigs that would perform well in our simple facilities. They both did this well, and often better than this. Well, the simple truth is that "all pigs is pork." We were rewarded by taking the time and making the investments to create Durocs and Chesters that would work for us and the people who bought their offspring from us.

The next consensus of the livestock panel was that at this point, the breeding philosophy for your foundation animals may be every bit as important as the breed type.

The buyer should seek out producers working with facilities as similar to his or hers as possible. The animals should be from a similar environment, be a middle-of-the-road type, and be bred for a balance between performance and reproductive traits. On this subject, we may have gotten a bit overly emotional, but the Hamp breeder made one last point that still sticks with me. He said to place a premium on a producer who was ". . . at least 50 years old and never goes to hog shows."

There is a time and a place for breeding animals that are rather extreme in type for one or two traits. They are like hot sauce in chili—a dash or two may be needed to tighten the whole thing. Sadly, it almost always seems that extreme genetic pieces come with a price.

Selection for Range Type

The hog for the out-of-doors will present a fairly distinctive appearance. These hogs will project vigor, soundness, and structure. They will need a bit of "cover" or "finish," which is industry-speak for the fat layer hogs deposit over the top-line, beneath the skin. Range hogs will also need good end-to-end and top-to-bottom width and body capacity.

The selection process is basic to the successful launch of a swine venture and is the one process that causes most folks great concern. Any animal can be picked to pieces if you take the time, and therein lies one of the keys to the selection process.

First, form a mental image of the animal that you want, the ideal for your particular set of circumstances. The animal you are considering buying can then be compared with that mental image. Animals from even the largest groups can then be singled out and fairly rapidly be matched up with this mental snapshot.

For the selection process, it is also wise to break down the hog you're considering into "pieces." The three main pieces are the feet and legs, the body, and the head. This concept is useful because it can also help to pinpoint areas that will need improvement as the herd progresses.

The place to begin is an area that is critical to the success of the range operation, the feet and legs. The foot should be large and well formed, with even toe points. Many farmers believe that a large foot is a good visual indicator of growth and future frame size. Many old hands hold that the young hog will "grow to meet a large foot and jaw."

Along with being large, well formed, and even, the toe points should be free of chipping, cracking, or any other signs of damage. Fine cracking in the hoof points may indicate biotin or other deficiencies. I don't believe I have ever seen a hog fully recover from a toe point injury. They are clearly quite painful and debilitating.

The hog should walk out in a rather flat-footed or "coon paw" manner. This assures long-term soundness, natural cushioning, and durability. When I was still is high school, swine production was swept by a fad for high-topped hogs that had long, straight front legs and that walked on their toe points. Fortunately, the animal-rights advocates weren't so much a factor back then, as those toe steppers were a problem from the get-go. Huge numbers of them developed bowed knees, and a troubling number of them became completely broke over at the knees.

A hog's feet and legs were meant, among other things, to serve as natural shock absorbers. With this in mind, you should select for long, sloping pasterns. Do not consider animals with stove-pipe-straight legs or those that show cocked or bowed knees.

It may seem like I'm stating the obvious, but for soundness sake, a hog should have a leg out on each corner. Hind legs that are tucked too far under the hog will cause soundness problems. Weed out hogs that show a scissors-kicking action in the hind legs as they run or trot. Scan all leg joints for knots and signs of swelling. On the back leg joints of some hogs, you might see knots or spurs from calcium buildup. These are termed "floor rubs" and are generated by contact with hard and unyielding concrete and slat surfaces. I've bought a few with modest floor rubs and used them successfully outside, but those hogs coming off of confinement floors have to learn to walk all over again.

The best view of the feet and legs comes when the hog is rising up or lying down. I like to view them doing both. When approached, a pen of hogs will naturally raise up and move away from the intruder. Give them a couple of minutes, and their natural curiosity will have them working their way back to you. Watch for those that walk about with a long stride and free and easy movement. Don't consider any that appear too tightly wound or walk with short, mincing steps.

Deep bedding in a sale ring can hide feet and leg problems. In fact, a deeply bedded sale ring should send up a few red flags. When I'm narrowing my choices, I like to work them a bit, and driving them slowly while snapping my fingers seems to energize them and get them moving and flexing quickly.

The body cavity of a good outdoors animal will be wide and deep. The best possible counter to respiratory ills is bred in lung capacity, and deep sides are indicative of growth there and the animal's ability to consume and convert large amounts of feedstuffs.

Good width between front legs and those in back—at least the width of a fully outstretched hand—is your guide for the chest area and underside (or "internal") dimensions. Those large lungs need a wide chest cavity, or "floor." Old farm hands say you should be able to roll a basketball between the front and back legs without ever touching the hog. I don't believe I ever saw one quite that good, but you sure don't want both of those front legs coming out of the same hole.

In the hindquarters of the hog we find the money cut, the ham. Over time, we have been told to breed for long hams, boxy hams, deep hams, round hams, and sometimes a combination of these. For a while, we had to chase hogs around the ring at judging competitions to hand-measure the diameter at the hocks. Hogs just don't appreciate that degree of familiarity from common strangers.

At the moment, extreme muscling is the fashion, and this is the second time we've gone down that road in my lifetime. Those tightly wound gilts had trouble farrowing, and many walked with the pitch and roll of a schooner rounding the Horn. The range producer will be best served by adopting a sort of puritanical approach to muscling: everything in moderation.

Personally, I like to see a large tail, with base of the tail (or "tail head") set well up on the body. As with jaw and foot size, a big tail is something the animal will grow into. The high-set tail head tends to draw the back legs out and is indicative of both good ham and carcass length.

The top-line should be long and level, with a bit of a turn out over the loin. Very heavily muscled animals will show a depression ahead of the tail head, which has come to be called a "meat dimple." This is created by exceptionally large and flaring hams. Down the center of the back of a heavily muscled animal will appear a rather pronounced groove where the loins emerge. Such animals are said to have a "sharp turn over the top."

Be watchful for a break or flat place in the top-line ahead of the hams or over the shoulders. In the long term, they can have an impact on soundness and durability. Likewise, a very high or steep arch in the back is to be avoided.

The shoulders produce the premium cut termed the "Boston butt" in many meat texts. From them come that wonderful staple of the Missouri barbecue season—which runs from roughly January 1st to December 31st—the pork steak. This part and the loin produce most of the pork roasts. Shoulder muscling should be proportionate to the rest of the animal.

Veteran swine producers have been known to wile away entire days discussing naught but the head type of a good hog. A head that appears too large for the body, very long eyelashes, and coarse hair can indicate an animal that has stalled out or become stunted. The head doesn't offer a lot in the way good eating, but it can tell you a great deal about the animal in general. For example, the wider set its eyes, the wider its internal dimensions are usually going to be.

The matter of jaw size has been discussed above, but it is just one of many head features producers look to for clues about the overall animal, such as early expressions of masculinity or femininity. A good

Deciding How Much to Pay

Early on in the selection process, you must resolve how much to pay for the animals you desire. The old rule of thumb is that a 40 pound feeder pig has the equivalent value of a hundredweight of butcher hog. It's a neat little formula, and it actually works a fair amount of the time.

It does not allow for that old tenet of supply and demand, nor does it offer any premium for quality. It does afford a base point, but with pigs in short supply (now most shoats are under contract from birth) or of exceptional quality, price adjustments can and should be made.

Feeder pigs bought for the out-of-doors should be no lighter in weight than 40 pounds, and in inclement weather a minimum of 50 pounds is even better. Shoats from a hot nursery have a very limited window to make the transition from life inside to out. The thinking is that it follows somewhere around 60 pounds. In cold weather, they represent simply too great a risk to take outside. Also, expect to pay more when buying just a few head of shoats.

In a relatively few years, we have seen butcher hogs sell for as little as 8¢ per pound, and a handful of boars sell in the six-figure range to head up artificial insemination studs selling to the show pig trade. A lot of good boar pigs have gone under the castration knife of late for that same show pig trade. Good young breeding boars can now be had for roughly double to triple the price of a single butcher hog. In point of fact, the $250 boar is on the way to becoming the standard for the whole of the boar trade.

Gilts now are the most desired commodity of the seedstock trade. Gilt pricing has always been a sort of will-of-the-wisp sort of thing. I don't think I've ever seen a formula advanced for pricing them. With so many hogs now under contract, I fear the days of buying a set of good F1 or F2 gilts from a nearby farmer/finisher are behind us. They won't be abundant, nor will they be cheap, so the best advice may be to start small and solidly and to plan on producing your own female herd replacements.

F1 gilt producer we know sets a high value on well-formed, large-diameter nostrils.

An old superstition among Hampshire breeders had them shun any animals with white tips on their noses. Sale catalogs used to denote each hog as "bn," "1/2 bn," etc. The "bn" was breeder shorthand for "black nose."

Selecting for Health

As important as knowing what a good animal looks like is knowing how a healthy one behaves. You certainly don't want to buy someone else's mistakes or any animals that they've given up on.

The places where most folks think of going to buy hog—sale barns—are no place for amateurs (or pros either, for that matter). I've sold a fair number of hogs through the sale barns of the Midwest, and I even bought a few there. They are, however, the marketplace for which the term "buyer beware" was coined.

I've been in sale barns that had a manure pack at least 50 years deep on the pen floors. Most are run like graveyards: they take all comers. I have seen hogs of excellent health and quality penned next to animals exhibiting all the signs of classic rhinitis. Hogs bought at such markets at the very least have to be considered to have been exposed to most parasites and to be fairly stressed.

Swine shows also present some unique problems for potential buyers. At one time, the major swine shows were essentially the showroom for breeder swine sales slated later in the year. Breeders would go so far as to charter livestock rail cars and travel from show to show by train, filling buying orders and selling breeding animals to local farmers. The hogs at a show are of first quality in most instances, but they have been brought together from many different farms. Some have traveled quite far, drinking different water and having gone through the stress of fitting and show-ring competition. They are under stress and are susceptible to illness or disease.

A buyer at one of these shows may have also traveled a long way and may have to make a purchase at a show. The constantly changing environment for the show hog can take its toll. If you prepare well, this is the time for some good old TLC. Use extra bedding, shield them from all drafts and other kinds of weather stress in transit, add vitamins and electrolytes to the drinking water, buy the same feed they have been receiving, and when home, make changes only gradually and pen them well away from the hogs.

The Extreme Animal Type

As you shop for foundation animals, you will often be offered animals that are very extreme in one or more type traits. For example, over the years we have produced boars that were exceptionally tall, long, or lean. These are animals created with a "purpose."

Farmers with herds in which there are glaring faults will need animals of extreme type. Such animals, however, may lack the balance or substance needed in other areas. They are used best very briefly in the breeding herd and then taken out. They are like that dash of hot sauce that is needed to bring the dish together. A breeding program can be quickly skewed or even broken down by the overuse of very extreme genetic pieces. Those thin-rind animals, for example, tend to lack vigor and durability.

Better still, if you see something you like, schedule a visit to the breeder's home farm, if at all possible. If that animal is part of a deep and well-grounded breeding program, there should be many more good ones back home, and they may even be priced a bit more modestly. If that is not the case, then that's something you should know before investing in those bloodlines.

The single best place to buy a hog is on its farm of origin. The benefits of selecting animals there are many:

1. On the farm, you can see how the animal was produced (it should come from facilities as similar to the ones you have at home as possible).

2. You can view full- and half-siblings to see the strength and uniformity of the genetics behind the animal.

3. You may even be able to view the breeding herd, including the sire and dam.

4. If it's a small farm, it's probable that the small farmer working with limited numbers must value genetic depth and consistency.

5. You can visit with and get to know the producer.

6. The stress load on the animal will be much less if taken directly the farm.

To determine the state of their health, you must apply the fullest possible focus to the animals you're considering purchasing. My father and I had a system that came together rather naturally. He would stand well back, observing the animals as a group, and individually, as the sum of their pieces. I was the up-close, hands-on part of the partnership.

I remember one spring when we decided to add a new Duroc gilt to the herd. I went early to the breeder's sale to walk all of the pens and handle the animals individually. Dad arrived later and from the opposite side of the barn caught my eye and pointed to a gilt in a large group that had just been turned out for exercise and water. She turned out to be the top gilt on both of our lists.

A quick health checklist should include the following: (1) free and easy movement, (2) a clean and bright hair coat, (3) clear and alert eyes, (4) no weepy discharge tracks from the corners of the eyes, (5) no labored breathing or nasal discharge, (6) no knots or swelling on the head or legs, (7) no signs of scouring (diarrhea), (8) no frequent sneezing, (9) no abscesses (10) no gauntness, (11) no listlessness, (12) not a narrow body, (13) no raw wounds, (14) lack of any soundness problems, and (15) an alert and vivacious manner.

Another important reason for buying hogs on the farm of origin is to get a real feel for the health and background of the animals. Every farm will have the occasional sick or lame animal—it is a given—and this should be expected in your visit. However, if two or three unsound animals emerge from every pen or house that's passed, if you hear respiratory ills even before you reach a pen, if lots are cluttered, groups are very uneven in size, and there is that general sense that things could and should be better, this is not the place to invest your hard-earned dollars and herd-building dreams.

I also like to take the time to get to know the producer and his or her philosophy of production. If you are taken on a tour of the operation in a pickup littered with vaccine bottles and syringes, buy there at your own peril. There are producers who are comfortable with an injection for every sniffle and twitch, and they carry syringes in their shirt pockets like most folks carry ballpoint pens.

I'm not anti-medicine and believe every producer should know how to give a variety of injections, but I also know my share of drug cheaters. Steroids will make many soundness problems go away temporarily (at least until the check clears), and hormones can have even barrows rattling the gates like yearling boars for a short period of time. A producer should be able to answer your questions about regular health practices, drugs and procedures used, feeding practices, and the timeframes for the different practices and treatments used.

I've known some stock traders over the years that made good money buying "rough" stock out of area sale barns and getting the critters back on track. They had old-fashioned good luck plus the skill levels that enabled them to know which animals were simply underfed or carrying a parasite load and those that had real health problems. Most of us aren't that smart or lucky; we need all the input possible to guide our decision-making processes. And you certainly don't want hogs for the out-of-doors that have been given a shot for every sneeze and sniffle.

Over the years, one of the selling points for our boars was a hardy constitution that came from a sort of benign neglect. As I used to tell prospective buyers, "If I can't kill 'em, you ought to do pretty well with 'em."

I wasn't really that hard on them, but in response to some very hard-earned lessons, I realized early on that it was very unfair to offer breeding stock that had been excessively fitted out and fed in an impractical way. What finally decided this for me was listening to one of the legendary breeders in the Yorkshire fraternity. His breed had long been termed "the" mother breed, and he had done this by resolving to make his herd live up to that standard. He fed no pre-starter creep feed or starter of any kind until his pigs were weaned. They made their start in life on mother's milk and what they could rob from their dam's feed pan and spills.

I've seen boars grown to 300 plus pounds on naught but pig starter. They were tightly penned, with limited exercise, and were handled more like Pomeranians than pigs. To put it bluntly, a quality boar is going to spend a good part of his life on his hind legs, and to that end he has to be fit and athletic. The range producer must have genetics that will work under the Sun and in the snow and always in a practical manner.

Selection for Reproductive Performance

The Sow

About 2 years into our breeding stock operation, I made one of the classic greenhorn mistakes. Our main product was boars, so I figured, why not breed from more masculine looking females?

With this in mind, I bought a big-headed, large-framed young sow that was really big in the front and was long. She was certainly durable, but that may have been due in part to the small size of most of her litters. Her litters were often quite uneven—the pigs sometimes seemed to be put together from odd parts—and she never produced the boar pigs we had envisioned from her.

My very first purebred swine purchase was a gilt, and I suspect this is the same for nearly everyone else. Gilt selection is one of the most basic of swine production tasks and is often the subject of great debate.

To invoke a rather poor analogy, the female is essentially the plant or factory that produces the crop of pigs. To that end, the female must be fecund, durable, and of reasonable meat type. The story is often told of the old Missouri farmer who picked his replacement gilts from the last few females left in the fattening pen each year. Like every other farmer, he needed income as fast as it could come available, but by choosing these lesser animals, he was selecting for poor growth and mothering performance in his female replacements.

A few old-timers of my acquaintance select replacement gilts only from their oldest sows. Their reasoning is that these females have had the time to build up the greatest levels of natural immunity to the "bug" mix on the home farm and pass it on to their offspring. Also, they are the most durable of females. Their performance has been proven dependable in the long term, and selecting them is in keeping with the family farm tradition of maintaining individual breeding lines that are proven to the home farm and the locale.

I've been known to work my way through a great many gilts, looking for the one or two we wanted to fill a need on our small farm. I'm not that hard to please, but every farm is different and really rather unique unto itself. Time and experience has shown what works for us here, as well as what our buyers expect from the animals purchased from us.

At a minimum, select gilts from the largest third of the gilts in a pen. They are the fastest growing and are the beneficiaries of that good early start that comes from good milking and mothering by their dams. If the one you select is from a gilt litter, there should be a minimum of eight pigs farrowed and eight weaned. From a sow litter, that number should be at least nine. The gilts should also have a minimum of 12 well-spaced and evenly formed teats, at least six on each side of the udder.

Many will boost those weaning average criteria by a full pig or more, but the argument can be made that as a producer, you are better served by having litters of eight to nine large, even pigs than by irregular litters of 12 to 14 or more. Our biggest litter to date is 18, and frankly, it was sort of a mess. A couple of the pigs weren't much bigger than a buck gerbil, and we had to tend her and her litter very carefully until weaning. She raised 11, and that hard start marked them for life.

I once owned a Chester White sow that regularly farrowed and raised seven big pigs from a gilt onward. It was a bit out of character for one of her breed, but they were always exceptional pigs, and for her I broke my own eight-pig rule. A littermate sister beat her average by three or four pigs, and hers were always among the fastest-growing on the farm. I kept her for several years and sold many of her sons that performed well for others. Still, that litter size was a sticking point for several potential buyers.

One key to keep in mind is to select a gilt that fits what might be called the "feminine profile." I noted above the importance of fast growth and frame size, but a gilt must still maintain a feminine appearance reflective of hormone function and her maternal role. A hog that appears feminine might sound like a bit of an oxymoron, but it is a selection must.

Example of a prominant underline.

A good female replacement will have that slight look of refinement. Even at a glance, gilts and boars should appear different in profile and stature. Some breeders go so far as to say a good gilt has a "milky" appearance. She will be deep-sided, with a long, smooth underline. Be wary of very tightly muscled gilts with underlines that are drawn up in the flank, because during lactation, they may have udder segments that become inaccessible to nursing pigs.

The teats on a good gilt should stand up so that they are well-formed, evenly spaced, and visible at a good distance. Cull for pin nipples, inverted nipples, or nipples too close or irregularly spaced. However, do not be too troubled by uneven teat counts such as six teats on one side of the underline and seven on the other. I've seen a few gilts, mostly Landrace, with 16 teats, and I owned a couple of Durocs with a 15 teat count.

Internal dimensions and leg soundness are also crucial in gilt selection. These animals need good body capacity to carry large litters and to consume sufficient amounts of feed to meet their needs and those of their pigs during lactation. Their legs must carry the sow through many lactations and seasons of mud and snow. A large, well-formed vulva is the only outward visual indicator that the female has a well-formed and developed reproductive tract. Avoid females with any knobby points on the end of the vagina or vulva.

I also prefer to pick gilts with multiple siblings that would be considered "keeper" quality. I believe this attests to the strength of the genetics behind the gilt.

The Boar

The boar is 50% of the herd, and good ones will move you ahead light years with their first pig crop, while others may turn out to be little more than sow fresheners. Two good boars used in succession can make up to an 85% improvement in most of the major economic traits.

You want a rugged, masculine, fast-growing boar; but again, be wary of the overly extreme individual. Has the animal achieved extreme performance in one area at the expense of equally important performance in other areas? Some farmers hold that the time to select boars is either on the day of birth or not until they weigh at least 300 pounds. Such is the changeable nature of a young and developing purebred boar.

There are numerous odd and awkward stages in a young boar's development. I can personally attest to the old adage that at least twice between birth and 300 pounds, you will have cause to scratch your head and wonder, "Why did I ever hold on to that pig?"

Also, don't forget that the greatest impact of a breeding boar will be seen indirectly, through the performance of his daughters and granddaughters, in what they give back to the breeding herd. The reproductive abilities of his dam and grandams are thus a very important part of the boar selection process.

For example, a boar's litter size at birth and weaning is every bit as important as that of female herd replacements. When selecting breeding stock, it is not always possible to view the breeding herd that produced them, but if at all possible, ask to do so when visiting a farm to select a herd boar. They will be in their "working clothes," rather than being show-ring slick, and if they are sound, well grown, and still in place and producing at good ages, it bodes well for his potential durability and performance.

A simple checklist for boar selection should observe the following:

1. Large frame and a ranking in the top quarter of the pen for size and growth (many breeders will not buy from a pen of young boars that has been picked through more than a time or two).
2. Large, well-formed testicles.
3. Large, well-formed and even hoof points.
4. Have at least 12 well-formed and evenly spaced teats on the underline (at least three teats ahead of the penile area is a very good visual indicator of carcass length).
5. Be of good breed type.
6. He should walk out with a long stride and great mobility.
7. He should appear vigorous and agile (do not confuse excessive aggressiveness for strong libido).
8. Have an even temper and willingness to be handled.
9. Display a true masculine appearance in outline and through the head.

Some boars are sold with various amounts of what has come to be called "performance data." This will generally cover the loin-eye size in square inches, back-fat cover, feed efficiency, and growth rate as determined by days of age to reach a certain weight. These are often termed the "economic traits" and are believed to have a higher degree of genetic inheritability than the reproductive traits. Experience has shown that data can be gathered or it can be manufactured, so you must always question just how realistically the data on an animal were obtained.

If you tightly pen a young and growing boar, feed a high-protein and very costly ration, slop-feed the boar (feed a gruel of feed and water), this can skew feed performance and growth rate quite dramatically. Most market hogs are fed out in fairly large groups. They have to compete at a self-feeder for a simple, meal-type ration, and the protein levels in these rations are often decreased as the hogs mature. The practice of pelleting may increase performance from feedstuffs by as much as 3% to 5%, and intact males have a hormonal stimulation for growth that barrows and gilts lack.

Back-fat is given as an average of three probes along the back or top-line of the hog. The argument embraced here, perhaps too firmly, is "the leaner the better." Pork, after all, has been fighting a fatness image for generations. Leaner is good, yes, but too lean may even be a greater problem.

Right now, you can see a lot of boars being offered with a back-fat average down around half an inch. These are boars that are bred for life inside a very structured and controlled environment, and even there they have not been without their problems. As a producer, this is the second time in my life I've noted that the swine industry has become overly enamored of the thin-rind type of hog. The first time the type pendulum swung this way, it was very quickly driven back to the other extreme, to get back to square one. Producers then were encouraged to select for "spongy" and "Jell-O middled" animals. The hog folks don't always get it right on type, but we are dead-on with terminology and colorful phrases.

A hog needs a modicum of fat to fuel hormone production naturally, to provide some natural insulation for a life lived out-of-doors, to add a bit of added flavor to the cuts of pork, and even to serve as a bit of natural cushioning against hard and unyielding surfaces. Fat also adds to the cooking qualities and presentation of pork cuts. The Japanese, one of the larger buyers of our pork exports, make a point of requesting pork from butcher hogs with at least an inch of back-fat.

Given the time, any bird or animal variety can be bred to extremes. They can be bred to standards far beyond the limits of practicality or economic justification. Massive muscling, for example, can later create problems with both soundness and reproductive performance.

It has been a long time since the hog resembled its wild ancestors. Still, we must never forget that that is the prototype from which we are to work as swine breeders and producers. Depart too severely from the traditional mold of an animal that can function freely and comfortably in an ever-changing outdoor environment, and you have a badly compromised creation.

Where performance data is available (it now seems to be offered much less often than even a few years ago), I tend to opt for middle-of-the-road levels of performance. There are those who argue that a boar that is very extreme in his type may give you a great leap forward within a single generation. This is possible, but it might also cost you many generations getting back to where you were when an extreme animal took you off the rails. If you are in need of big changes, even a middle-of-the-road boar will move you forward. As with the gilts, I prefer a boar from a litter that had at least two or three boar pigs that were of "keeper" quality. To me, that really signifies depth in a mating.

With each new boar, you can realistically address no more than one or two weak areas in the breeding herd. Here, you use the old rule of thumb of breeding strengths to weaknesses. If loin-eyes are small in the butchers that are being produced, select a boar with a good-sized loin but that still retains a good balance for all major traits.

Thus, my numbers criteria for boar selection are an inch or a bit less of back-fat, loin-eye in the $5^{1}/_{2}$ to $7^{1}/_{2}$ square inch range, and a feed efficiency of around $4^{1}/_{2}$ to 1. (That is, $4^{1}/_{2}$ pounds of feed to produce 1 pound of animal gain on average. This ratio grows as the animal matures, and generally, it will take more feedstuffs to produce fat than lean matter.) And never buy just according to the numbers. Get to know as much as you can about the animal, like how and what the growing boars have been fed, for example. If they have been fed out in pens of five or less and were fed pig starter all of their lives, there is simply no validity to their performance data.

Boar selection should be guided by your boar needs—sounds so simple, but sometimes quite hard to do. It begins at home with a clear, hard look at both the sow herd and the pigs that they have been producing. The good hog will produce animals as good as or better than itself. And the good herd is thus always moving forward.

The Breeding Arts

The two classic breeding plans that have been followed by stock raisers since times long past are to selectively breed genetic strengths to genetic strengths or to selectively breed to upgrade one or two of the weaker areas in the existing herd and their offspring. To see these flaws, you must overcome a tendency toward "barn blindness," or self-deception, and see your charges as they really are. Many farmers have to seek outside help with this task. You can get good feedback from extension staff, veterinarians, or some feed company field reps. Once upon a time, the major packing firms had people in the field to help with animal evaluation and selection.

My father was of the old school, and he kept our herd notes in pocket notebooks from seed and feed companies and on a big old kitchen calendar we got each year from the local bank. With simple notes made in the field, he was able to track such important matters as mothering, growth response, feeding, health care, and reproductive performance. He had that gift of all good stockmen, the stockman's eye. He had long ago formed mental images of the animal types best suited to his goals and means and then developed senses and skills that enabled him to quickly focus both on quality and potential problems.

Those goals were always realistic, and there was never a quest for some mythic ideal that never could really exist. Nor were these images and goals static. Change is inevitable, so there must always be some elasticity in your selection process and criteria.

When selecting replacement gilts for our Duroc operation, experience taught us that the two traits we always had to emphasize for optimum performance were body length and litter size. Those two always seemed the first to slide, even if neglected only slightly. With our Spots, leg structure was always critical, and with our Chester Whites, we could never ease up on growth and muscling.

Our buying trips were then nearly always intended to fulfill a need determined by careful study of our records and an honest appraisal of what we had going on back home. One time we needed a Duroc boar not for massive muscling, but for better muscle definition. Our genetic needs were for a bit better turn over the top and a better-formed ham carried farther down on the leg. Experience had taught us that we couldn't compromise on litter size or length in the selection process, so the task was not to form an image of the best Duroc boar, but the best Duroc boar for us. He would be from a litter of at least nine born and nine weaned, be a fairly "big outline" boar, and carry a nicely formed loin and ham, in balance with his body. With such a boar, we could key in our two greatest needs at the moment, without too much worry about something else slipping.

There are also a few ways to increase the bang from your breeding stock buck. Young boars can sometimes be bought for less than service-ready animals. If you have the time and facilities, you can get them into the herd and develop them as you desire. If you are trying to bring more consistency to your pig crop and use multiple boars, opt for full or half-siblings whenever you can buy good quality animals. If the two best boars you are being offered come at a good price but aren't related, I wouldn't hold it against them, however.

Good boars and gilts aren't as commonly available as they were even a short time ago, and especially if you're hoping to work with some of the heirloom breeds. Now more than ever, you must allow plenty of time for the selection process. And you may have to reach ever farther afield to find the needed genetic pieces.

I once helped my wife's cousin assemble a group of 10 Hamp-based crossbred gilts to start up a commercial breeding herd producing market hogs. A couple of phone calls later, we were given access to one of the better commercial swine herds in the state. I was a little concerned going in, as he was absolutely adamant about two things: (1) they had to be blue-rumped Hampshire X Yorkshire gilts, and (2) we had to come away with 10 of them.

The farmer/breeder graciously offered us access to his four largest pens of market-weight animals. The best animals before us that day were the Hampshire X Duroc females with their distinctive Black or Hamp color patterns. Two hours later and after working across several hundred head, we had nine pretty good blue-rumped gilts penned in the adjoining alley. That tenth gilt was a compromise at best, and she never performed up to the averages set by the other nine.

When Phyllis and I began our Chester White herd, we actually bought the herd boar first (when a good one comes along at the right price, you buy it). Our goal then was to match him up with eight good gilts. At the time, we had a good Chester sow at home that had gotten us some good F1 pigs when crossed with colored boars, so we went to her home farm to make our new selections.

The breeder worked with a very small herd, by the standards of the day, but he drove out a real good set of gilts. Six were fairly easy picks and were all similar in their breeding. All were long-time "dirt" hogs. Still, it was clear we were not going to get eight gilts there, and the seventh was a real decision. One of our first picks was a littermate of the last gilt we had picked, and the breeder had high praise for the gilt in question. In fact, he was keeping a littermate that was similar to her in type.

She was just so very extreme in some of her traits. She was quite lean, very long for a Chester, and even a bit angular in her appearance. We bought her, and she went on to produce some very interesting hogs. Her litters were small by Chester standards, although quite respectable, and she produced many long, trim sons. They were not as durable as thicker boars, and one proved to be a spotty breeder, although it produced good butcher animals.

You breed good hogs by assembling genetic pieces and blending them into the herd with a hand made deft by experience and study. It is not so much an endless task, but rather a continual challenge to create something better.

The range producer will blend genetics to form hogs that are both practical to produce and pleasing to the end consumer. That is indeed a rare thing in modern production agriculture. This is the creative process at its best.

Feeder Pig Selection

I've often thought that the greatest challenge for the stockman's eye is the selection of good feeder stock. You have to be able to see animals not just for what they are, but for what they will become.

I was raised a bit of a sale-barn rat and have seen literally thousands of feeder pigs bought and sold. Some of those animals would be described by those ever-poetic Midwest auctioneers as "being fresh as spring water" or "bright as a new penny." Others, however, might be described as needing "just a little extra feed and straw." The latter could mean anything from needing a good worming to "dead by dark." So buyers remember: Ballet dancers aren't the only ones who need to be on their toes.

A good feeder pig is simply not a scaled-down version of a good butcher hog. And you most certainly want to avoid buying somebody else's mistakes.

Most buyers like to acquire feeder pigs in large droves from a single source. Ideally, that source is not the back of a pig trader's truck. Buyers are after consistency in the end product, as well as a group of pigs that will move through the growing and finishing facilities in fairly tight order. And as always, the best place to buy them is on the farm of origin.

Feeder pigs are generally sold in the weight range of 35 to 70 pounds, although I have seen a few heavier pigs being offered lately. The old rule of thumb is that you buy lightweight pigs in good weather and heavier pigs (50 pounds and above) in cold or damp weather. Pigs much above 70 pounds are often considered suspect by some veteran buyers, the suspicion being that they might be the castoffs or the last of a pen of pigs that have simply outgrown them. Some buyers also want a pig with some color on the hide when the wind blows and the snow falls.

This last point raises the question of what a feeder pig's breeding should and shouldn't be. Yes, all pigs produce pork chops, but there are some valid breed considerations when considering feeder pig choices. White pigs, for example, seem to tolerate heat better than dark-colored animals, although they are more prone to sunburn. Black and dark red animals, on the other hand, have long been believed to handle cold weather better. This has been reinforced, at least in part, I believe, by the fact that most of the pigs coming out of confinement carry substantial amounts of White breeding.

A crossbred pig is favored over a purebred for finishing, due in large measure to the vigor and early boost in life it received due to heterosis or hybrid vigor. "Crossbred," however, means from a planned program of crossbreeding and the random throwing together of animals, which might more correctly be termed "mongrelization." Most crossbred pigs will be the product of a three-breed rotation–Black/White/Red—and thus maintain a degree of heterosis in the 85% gradient.

The advantages of a well thought out piece of crossbreeding are quickly apparent, since F1 and F2 pigs are more vigorous at birth than their purebred counterparts, with heavier birth weights and in larger litter sizes than those of their parents. That larger size at birth and the early vigor will be reflected throughout the animal's life.

What a good feeder pig should first be able to demonstrate is great good health. An alert manner, bright eyes, a clean and shiny hair coat, good size for its age, and soundness: all are essential traits to look for in the selection process. Watch closely for signs of respiratory ailments, such as coughing, sneezing, labored breathing (sometimes called thumping), nasal discharge, and swollen or twisted snouts, as well as indications of scouring, sunken or clouded eyes, a head appearing too large for the body, or excessive gauntness. Good pigs will be very uniform in appearance, sorted closely for size and quality. They will be free of internal and external parasites and typically well past the stress of weaning, and the males that have been castrated will be well healed. They should be ready to go right to the feeder to begin making money for you.

A good feeder pig should show a potential for rapid, marketable growth. That means they should possess a frame with flex and natural extension, deep wide chests, deep sides, good length, and a developing muscle pattern. I don't like to see young pigs too tightly wound in their musculature or that appear excessively tall or finely built. They should be solid to the touch, have growth potential, and be lively and curious about their environment.

My dad was an earnest marketer and promoter. I once left him behind to pen a group of our pigs at a sale while I parked the truck. It took me a few minutes to get back to him, and by then he was into marketing mode. Several folks were standing by the pen, including Dad, who shook my hand and inquired if I might be the owner of such fine animals. His role of very interested potential bidder might have played out better if the pigs weren't so busy rooting around the toes of his boots and pulling on those all-too-familiar pant cuffs.

As a buyer, you want feeder pigs "as bright as a silver nickel in a pail of spring water," and you must develop the skills needed to know the good ones, the pigs with real potential, when you see them.

Timing the Purchase

The ideal time to make a start with a livestock venture is during a down market for that particular species. You buy in as close to a market bottom as possible and then have production ready to sell as the market moves back upward. Unfortunately, there are no sure guides as to when to do that.

I have seen too many farmers make the mistake of trying to time the markets too narrowly to their own wishes. They will set a certain exact price at which they will buy or sell and will hang there even when the market begins breaking away just above or below their set point. As humans, we have a tendency to build these largely psychological barriers at very specific, rounded off numbers. A lot of folks holding to sell butchers at 45¢ per pound have seen markets climb to 44.75¢ and then begin falling back dramatically, for example. And on the buying end, I wouldn't risk not picking up a much-needed breeding animal for the sake of one more bid of $25. You may end up spending more than that in gas money trying to save a few dollars.

There is an old rule of thumb that says that you can generally buy swine breeding stock from fall-farrowed litters for less than animals from spring-farrowed litters. This and a few other old choice bromides have some truth to them. Still, I am a firm believer in making the plunge when you have made up your mind to get in for the long haul and have readied your facilities and operational plans.

One thing to always remember is that an animal as big as a hog should never be an impulse purchase. The impact of a single animal can be felt throughout the lifetime of the herd if that animal is either very good or very bad.

Form a buying plan and allow yourself time to track down the kind of animal you need for your particular operation. Invest some time and effort to become current on the swine scene; try to take the pulse of the hog market both in the short term and a few years out.

Prior to purchasing the animals, list the producers you wish to visit, phone ahead for appointments, formulate your price range, and make that all-important personal commitment to quality over quantity. Most of the great livestock herds and flocks can trace their roots back to a single, strong individual—most often a very strong female.

Too often in the startup stages, you can easily become the not-so-proud owner of less-than-optimal animals. Always seek out the very best individuals you can afford, and sacrifice numbers before quality. But be realistic in your pursuits. Sometimes you will encounter few spectacularly priced individuals, but if a producer's animals are priced

Transporting Hogs

Hogs may appear armored-tank tough, but some care must still be applied when transporting them from place to place. Heat, chilling drafts, and direct sunlight all can subject the animals to stress and can trigger any number of health-related problems.

The optimum vehicle for transporting hogs will either be completely covered with a roof or tarp or have sides that are high enough to discourage the animals from trying to climb up or jump out. Don't haul the animals on rainy days or during extreme cold. In cold weather, wrap a tarp around the front of the racks to keep chilling winds off of the hogs.

In warm weather, haul either early in the morning or late in the day. Including a few inches of damp sand on the floor to help keep the animals cool is a good idea. In cold weather, use 6 to 10 inches of dry straw for bedding during transport.

Drive carefully, avoiding sudden starts and stops, and keep a close eye on the animals during transport. Don't pack the animals too tightly, particularly in hot weather. Nor should you haul animals of different sizes and ages together in a group.

Have facilities at home ready to hold the new arrivals well away from any other hogs on the farm. They should go through a period of isolation and observation lasting at least 30 days. This will also give the animals time to adjust to their new environment and to your routine. Even a change in drinking water can throw an animal off its stride at this vulnerable time.

beyond your means, state that fact honestly. There will be other days, and you will nearly always be welcomed back.

On your buying trips, have the truck or trailer thoroughly cleaned and disinfected ahead of time. Arrive for your appointments on time, and avoid wearing clothing or footwear you regularly wear around your stock back home. Ask plenty of questions about feedstuffs, management practices, and health care the animals have been given.

Making Sense of Type Trends

As noted above, there have been many wide swings in swine type over the years. The explanations for this abound, but the one that rings truest for me came from an old gentleman with 50 plus years of experience working with purebred livestock. It was his contention that as typical representatives of a currently preferred type became widely dispersed and established in the industry, a certain core group of dominant breeders then needed something "new" to sell. The next new thing would give them a product bolstered by a demand that many folks might consider to be shaped by a certain amount of artifice and self-promoting hype.

There is some historical and chartable validity to this argument. The livestock industry has functioned largely on a pyramid-based structure. At its base and making up the greatest number of producers are those farmers raising butcher and/or feeder stock. The next level is a group termed "multiplier breeders." They are the ones supplying breeding stock to the base group that is primarily producing meat animals. At the top of the pyramid is a very small group, supposedly no more than 2% to 5% of the total number of livestock producers, who constitute the experimenters and type shapers. Their innovations are supposed to be based on the genuine needs of the meat trade, which are set by production circumstances and consumer demand.

Some of the type trends trotted out in my lifetime were very short lived, and generally for very good reasons. Some were actually attempts at a complete remake of the hog, and the results were often near-disastrous. Every type change shook out a few more breeders, caused others to be doubtful, often adversely impacting the meat trade and helping to pave the way for the seedstock companies. Hogs should be allowed to live, grow, and produce live young in fair numbers. Much beyond that, and you've got what amounts to genetic tampering fueled by little more than pure ego.

At the moment, the purebred trade is banking heavily—too heavily—on the show-pig trade. They are producing some of the most extremely muscled and leanest animals I have ever seen. They take as much tinkering and tweaking as an English sports car.

The range producer, however, must place his or her emphasis solidly on can-do animals. Vigor, durability, soundness, and performance need to be rounded into a single type package that embodies that "root hog or die" swine character of old. The good range hog is far removed from those razor-backed hogs of long ago, in so many ways. However, they do have in common a certain amount of grittiness, that natural strength to perform in the presence of an ever-varying environment.

Meat type should, of course, be a concern for each swine producer, but the animal has to be in realistic proportions and have marketable worth. One of the consequences of breeding for extreme type has been an overall decline in pork quality and flavor. Confinement production can produce rafts of so-so pork as a basic commodity, but pork as a meat of distinction and with savoriness that appeals to all the senses is quite another matter.

The packing industry has also done quite a disservice to pork as a preferred meat product. They have expressed a need for certain carcass qualities that do naught but improve hourly output from the packing lines. They have even restricted their purchases of animals of any other type, but they have never offered a truly meaningful premium for the type they say they need. The premium seen out at the country buying points has been for production in volume, and even then it has been quite modest. It was once a given that every buying station in the Midwest had 25¢ to 50¢ per hundredweight for that handful of producers that could deliver in high volumes.

The packers are also the ones who present pork chops and steaks with all of those fatty and silver-dollar-sized numbers hidden away in the bottom of the package. Most of that pale, watery pork from super-lean hogs cooks up dry and tough. Hiding the fat preys upon a long-held fear of pork as a fatty meat, which is largely undeserved.

Pork, good pork, is one of the most flavorsome and succulent of all meat choices. It has a corner on the American breakfast, it's a centerpiece at Christmas and Easter meals the way turkey anchors Thanksgiving, it's a top meat choice for roasting and barbecuing, and is leaner by half what it was at the end of World War II. A 3 ounce cut of pork is now the near-equivalent of a 3 ounce serving of chicken. In no way are range producers turning back the clock to the days of the lard hog; rather, they are creating once again the meat animal the consuming public has always valued and wants to have available to them again.

Pork, once the feast of kings and aristocrats, has suffered mightily not in the name of improvement, but of production efficiency. Even veteran producers now report they're unable to eat the pork being churned out of confinement facilities, which have a constant flow of cookie-cutter pigs coming out of one end and a river of wastes out the other.

In commodity pork, there is a creeping trend toward a generation of meat strip and nugget products similar to those now coming from the commodity poultry trade. We are starting to see this in mock ribs, certain patty products, round bacon for the fast food trade, and flaked, restructured pork "chops." The makers of these sorts of manufactured meats simply need a base flow of raw mass from which to chip, flake, flavor, dye, and restructure the meat to form the products they wish to market. Confinement production is thus the bridge from "farming" to forming.

The flavor and texture of these products is generally further modified with sauces, breadings, and seasonings intended to create some semblance of traditional or ethnic pork cuts and/or dishes. Each change greatly increases the selling price to the consumer, but effectively reduces all pork to heaps of product potential, to be haggled over for pennies. As one sale-barn vet put it, the packers won't be happy with farmers until and unless we can deliver hogs that are rolled out onto their docks in the form of skinless, boneless lumps.

Several years ago, a neighbor made a rather impressive point to me and everyone else within earshot when he claimed that the boars he bought from me were used to produce the pork for his family to eat, and that he used a very different kind of boar to produce hogs for the local buying station. In point of fact, a number of swine breeds have now been identified as producing exceptional quality pork. These breeds include the Berkshire, Chester White, Duroc, and Tamworth.

Berkshires or any animals with at least 50% Berkshire breeding produce meat that has come to be known as "black pork." This is a term borrowed from the Asian trade, where pork from well-finished animals (that is, a minimum of 1 inch of back fat) is highly valued. In fact, its meat is somewhat dark in color, and it also has interior flecks of fat that can be likened to the marbling in prime beef. This pork has a favorable texture or sheen and is different from others on many levels, right down to the pH values.

Tamworth and part-Tamworth animals are currently being marketed through a Midwestern-based program that is built around the unique character of Tamworth pork.

Taste is a function of many things, including visual appeal, aroma, and texture. There is much truth to the old bromide which holds that you sell steaks with the sizzle in the skillet and the aroma it creates. My grandmother's holiday hams, with their fruit garnishes, were works of art, and they gave holidays a special scent that met you at the front door. Memories of that sort, and the knowledge that there are conscientious farmers producing meat in sync with consumer values, also manage to make the food that much better and more valued by the consumer.

The range-reared hog, because of its improved muscle tone, varied diet, regular exposure to sunlight, and less stressful rearing environment, produces a very desirable pork for the family dinner table. Many of the new generation of producer cooperatives are organizing behind this concept of better pork through traditional genetics and more natural and humane rearing practices. What a unique concept: pork that's meant to be eaten and savored for the unique food that it is, rather than as another pale, flavorless commodity that's processed as quickly and cheaply as technology permits.

Crossbreeding

Most of the contract pork producers are now using composite breeding stock sold or supplied by a contracting company or an allied firm. These are generally crossbreeds, with substantial amounts of Yorkshire, Large White (the English York), or Landrace genetics. They're also the breeds with the longest histories of production in confinement systems. Some Danish Landrace lines are said to have been in confinement for over a hundred generations. To the composites are also added dashes of Hampshire, Duroc, or even a bit of Pietrain breeding for muscling and growth.

These composites have to be structured in just the right order or sequence to maintain some degree of hybrid vigor in the offspring. Any deviation from the proper sequence, and breeding can be quickly skewed, generally toward one of the White breeds noted above. Growth and/or meat type will be affected first, and as these are basically terminal crosses, there can be substantial economic consequences. Working with these crossbreeds also locks the producer to one more outside supplier and effectively cuts him or her off from that most basic of production needs, the seedstock.

All pigs are pork, and our surplus Duroc pigs always found a ready market as feeders among area farmers/feeders. Their early rearing had been simple and basic, and they performed well in the wooded feeding lots that are fairly common in our part of Missouri. Still, most independent producers opt for a crossbreeding rotation using multiple pure breeds to combine traits that are best suited for a single farm and its production system. This ability to naturally select for an animal according to one's needs and markets underlines and enforces one of the quintessential responsibilities of the independent producer.

The classic mixture is a three-breed Black/White/Red rotation. A prime example of this is purebred Yorkshire gilts bought for breeding to purebred Hampshire boars. The resulting female offspring that are retained as female herd replacements would be bred to a purebred Duroc boar. The resultant gilts would then be bred to a purebred Yorkshire male, and the three-breed rotation would begin again. With a three-breed rotation in place, an 85% degree of heterosis will be maintained in all offspring produced.

Of course this level of heterosis would be increased with a four-breed rotation, but it becomes that much more complex, and in a great many areas it would now be very difficult to find dependable suppliers of four different pure breeds. On the other hand, even with a simple two-breed rotation, a fair amount of hybrid vigor would be maintained in the offspring. This cross would be based on one White breed and one of the colored breeds. You would have to select for a bit more muscling and growth in the White boars, plus you'd have to be extra mindful of mothering performance on the colored side of the mating.

A few years ago, a great many breeders (myself included) jumped at the marketing of a supposed genetic shortcut, the F1 boar. These were largely Hampshire X Duroc boars that were supposed to combine the carcass traits of the Hamp with the ruggedness of the Duroc all in one package. Seen to a lesser extent were Hamp or Duroc X Yorkshire males that were supposed to add a bit of muscling while bolstering mothering performance. We kind of whistled past that old genetic truth that when traits are fixed to such a degree that they will breed true, you have created a new breed.

I sold a lot of pretty hot, trendy-type (or "typey") Duroc X Hamp boars and defended it to myself by saying that it was something my customers wanted and would buy elsewhere anyway.

And let's be very honest here. They represented something else I could sell without taking on the extra work of adding another entire breed. With just one breed, our customers would be back only every 18 to 36 months to pick that breed back up for their rotation. With the F1's, we were offering a poor second choice to the Hamp male in a breed rotation, but were making a sale we would otherwise not have.

The true advantage to an F1 cross—hybrid vigor—actually rewarded us as initial producers more than the buyer. These were quite vigorous pigs born in good numbers. The Hamp sow we were breeding to our Duroc boars never produced less than three keeper boars per litter. They went on to be fast-growing animals and vigorous breeders, but they lacked the genetic consistency in their offspring that could be expected with a purebred sire. They were an alternative that could be used in terminal crosses producing animals for slaughter, but it showed you simply cannot pack the best traits of two different, distinct breeds into a single animal.

Modern range producers will continue to be well served by a crossbreeding rotation when producing either feeder pigs or butcher stock. However, they may wish to draw on the more distinctive heritage breeds to capitalize on their more vigorous natures. There are numerous choices for the White, or "mothering," segment from the cross, and these include the Yorkshire, Large White (the parent breed of the York from England), Landrace, or Chester White.

For the Red segment, which generally contributes to growth and vigor, the choices include the Duroc, Tamworth, and Hereford. The Black sector adds growth and carcass quality, and choices there include the Hampshire, Spot, Berkshire, and Black Poland.

There are other choices, of course, but they currently exist in quite small numbers, and some of them have not been kept up to the standards of the past. Still to be found in a few places are the Mulefoot,

English Black, and the Red Waddle or Wattle. Not every Red hog with wattles is a Red Wattle, however. There are also some largely feral strains, such as the Ossabaw, which might be encountered in a few areas, as well as the Guinea Hog. When that market tanked, a lot of folks found the "Potbellied Pig" to be quite edible, but they are certainly not something you would want to build a breeding program around.

Purebred hogs offer the independent producer the fullest possible control of genetic resources. I grew up in what may later be known as the "second golden age" of purebred livestock. The quarter-century from 1960 to 1985 saw a vast number of modest-sized purebred herds supplying boars and gilts to what were termed "commercial hog producers," the broad-based group of producers marketing feeder pigs and butcher stock. It was farmers selling to farmers, and it was a market that flourished through the feedback and input from both sectors.

I could pick up a phone and buy a boar or gilt, sight unseen, three states away and know that what I requested over the phone would be delivered to my farmyard a day or two later. When I graduated from high school in 1968, the going rate for a good purebred boar was $85. Twenty-two of them would buy you a new, plain-Jane pickup.

Gradually, that price standard moved up to around $300, and it took 100 of them to buy a pickup. Needless to say, I long ago learned to discount any argument against the continuing role of inflation in the U.S. economy. Granted, both pickups and boars got a bit better over the years, but about the time the boar companies came along and started marketing swine genetics in catalogs instead of on farms, the swine world started coming off the rails of reality and good economics.

There were (and still are) farmers that would produce butchers and feeders all from a single pure breed, and with great success. We never failed to find a buyer for our surplus Spot, Duroc, and Chester pigs when offered as feeders. With a wide variety of pure breeds still available, it is possible to produce a crossbreeding rotation well tailored to your individual farm needs. And it is achieved with money that is generally spent quite close to home and that stays in the rural economy.

Range producers, for example, might wish to follow the classic Black/White/Red rotation, but with the special touches that are valuable for range production. There are some types of animal that just live and flourish in the dirt better than others.

The White foundation females could be Chester Whites, since they are known as the "durable" and "outside" White breed. Studies have shown that Chesters are the ideal White breed for crossbreeding to bring out the greatest strengths of the colored breeds while still expressing good mothering traits. The Hampshire X Chester pig has unfairly stood in the shadows of the Hamp X York for too long. This cross has won some mighty tough market hog shows over the years, and these F1 gilts can be real head turners. The only gilts and sows we ever had that could produce three litters in a 12 month period (with 5 week weaning) were Chester Whites.

Cross those Chester females with a Berkshire boar to impart meat quality and growth, and you are well on the way to creating a very distinctive line of range-produced pork. The Red breed could then be represented by a Tamworth or Hereford to continue in the heritage vein, but I would go with a good Duroc pig. The growth and muscling of the animals will be reinforced in the mix, and since Durocs are developing a modest reputation of their own for exceptionally flavorsome pork, they will be easier to find.

The Pure Breeds

In order to guide you in the selection of breeders for your farm, I've provided below a few thumbnail sketches of the purebred animals now available in the United States. We have raised several types, owned

YORKSHIRE

crossbred animals drawn from several of them, and nearly all have been raised with great success in our part of the Midwest.

The Whites

Yorkshire

Of the White breeds, the Yorkshire may represent the greatest option for growth and muscling. They are bred here in greater numbers than any of the other White breeds. Many years ago, I read a good article about a Western butcher-hog producer who worked with nothing but purebred Yorks. At the time, he made a strong case by claiming that with good selection, the York could very well be a top contender for best all-around hog.

The color for which these animals derive their group's designation is not exactly true white, but rather the palest of light browns. The Yorkshire has erect ears and good length; it's noted for foot and leg quality; and some lines have much better body capacity than others. You won't sacrifice much in the way of carcass quality with a Yorkshire, but be careful to select for a broad base along the bottom of the body cavity, from end to end.

English White

The Large English White, sometimes seen in the United States, is virtually identical to the York in type and conformation, and shares some space in herd books with the York. It is most often seen as one of the breeding components offered in commercial breeding composites, and has been bred in confinement for literally scores of generations. The breed has produced some very large-framed specimens, and some Yorkshire breeders put a strong emphasis on the English breeding in their lines.

Many swine breeds commonly seen in the Midwest were developed in Great Britain. All of the "shire" breeds hail from there. Many breeds continue to be cultivated there that exist only in limited numbers in the United States or Canada. The English White was actually said to have been bred in three different sizes, based in part upon the growing conditions to be found in the immediate area. There is also a Large English Black that now has a following in the royal family there. It reminds me of a Black Landrace or Wessex Saddleback, and the last ones I know of in North America were available from a seedstock firm in Canada, although I have not seen them advertised for several years.

Landrace

The Landrace is a very long-bodied White breed with the big, drooping ears. They were the base breed for the confinement industry in Scandinavia and Europe and are used largely in that role here. Muscling has been somewhat improved here, but their primary role is to increase litter size and add length. They have also been bred for a smaller ear and more substance in the head area.

The Landrace are a most docile breed, and this can put them at some disadvantage in group situations with hogs of other breeding. They can be easily cowed and pushed back from feeders. A neighbor of ours raised Berkshires and crossbreds to sell for butchers. Several years ago, he bought a group of high-percentage Landrace gilts and, out of necessity, had to drop a couple of younger Berk gilts into the pen with them. The two younger gilts came to dominate the group to such an extent that some of the White gilts rapidly lost condition (vigor and muscle mass) even when being offered a flushing ration (which is a full feeding regimen—3% of body weight daily—for sows just prior to breeding, for at least 2 weeks).

Many swine producers have reported problems with certain Landrace lines out-of-doors and in simple housing. But I can remember that when I was coming up, some of the more rugged old sows in the

CHESTER WHITE

country showed strong traces of Landrace breeding. Those farmers seeking Landrace breeding stock these days might be best served by using breeding lines that were present in this country pre-1970, particularly animals from farms where at least the sow herd has continued to be maintained outside.

Chester White

The Chester White is a breed that has always had a special place in my heart. They are a White breed with modestly drooping ears. They grow to a good size, and along with the Durocs, are quite frequent competitors in the Largest Boar class at the few state fairs where such competitions still occur. The first purebred sow we ever owned was a Chester, and she handled herself well and always maintained condition in the group-penned sow herd.

Chester pigs seem to be a bit smaller than others at birth, but this is in part because they tend to be born in quite substantial numbers. Some may even grow a little slower, but I always found the performance of our Chesters to be comparable to that of our Durocs.

I have elaborated on the Chester White above, but I should reiterate here just how compatible they are for crossing with the colored breeds. The range producer hoping to balance mothering and performance will find this type an optimal choice. Chester Whites are also a gentle-natured variety that we found to be very easy to work with in our one-sow houses. We often tended to pigs right there in the house with the sow.

We once had a set of Chesters get out, and the sows went one way and the old boar found shade about 20 feet from the pen and settled in there. Phyllis's younger sister Cathy offered to lend a hand, and we left her to keep track of the boar and then headed off down the road. Forty-five minutes later, we had all of the girls back home and the fence fixed, and Cathy and the boar were still right where we left them.

The Reds

The Red breeds are the hardy breeds, and they're also among the prettiest of the hogs. The three breeds in this group are very distinct. They all have a long history and a strong connection to the family farm.

Duroc

The Duroc is an American developed breed that started in the state of New Jersey. One of the breed's founders also owned the famed racing stallion Duroc, and hence this Red breed's original name, the Duroc Jersey.

We did 30 plus years with Durocs and found them to be consistent sellers; they were also popular with commercial producers for their ruggedness and good meat type. They have modestly sized drooping ears and hides in various shades of red, from a deep, almost plum colored hue, to a very light shade of red. From a distance, some old lines looked almost black, with slight yellow/gold touches.

Durocs are generally quite rugged in their type, and they're well known for their feet and leg strength. Lately, they have been bred for greater overall length. They are kind of a complete package when it comes to hardiness and muscling and are heavily relied upon for this reason by commercial producers. They are far and away the most widely available of the Red breeds and may have one of the deepest U.S. gene pools of any pure breed.

Tamworth

The Tamworth is a light red breed with erect ears. They were one of the breeds that were once termed "bacon" hogs and have a very long history for leanness. They are also the Red "mother breed."

TAMWORTH SOW

Tamworths milk well and are quite protective and responsive mothers. The nursing sows have a distinctive way of lying down that helps to prevent pig loss due to overlay. They first drop to their front knees and then sidle down slowly, scooting the pigs out and away from their descending bodies. They have a largely undeserved reputation for temper that may be due in part to the fact that many of the descendants of this breed have been left largely to their own ways for so many years.

They were the brush hogs of my youth, and some of the old-timers kept them with little more than a "feed 'em and forget 'em" style. This old breed has always been sustained by a strong breed group and never really slipped in the way that so many other old line breeds did. People who used Tamworths once have always seemed to go to the extra effort to use them again.

The Tamworth is another breed that is building on the quality and taste of the pork it produces. They have also figured quite prominently in a couple of "outdoor composites" trotted out over the last few years. These animals are perhaps the most modern of the heirloom breeds in their type and continue to have a strong presence at a few major swine competitions, such as the Illinois State Fair.

Hereford

The Hereford hog shares the distinctive red and white color pattern of the Hereford cattle breed. They also have a modest drooping ear.

Like the Tamworth, this breed also has the backing of a strong breeder group, and there is even an annual type conference and auction of breeding animals. Herefords have often been promoted as the world's most beautiful hog. At times, they may be a bit smaller than some of the other breeds, but they are quite competitive on the rail.

The Hereford is a breed that I believe is really poised for a move upward. The distinctive color pattern should lend itself quite well to marketing programs; I have seen some Herefords compete very successfully in market hog shows right here in the Corn Belt.

The Blacks

The Black breed group includes the Hampshire, Berkshire, Spotted, and Black Poland breeds. These are sometimes termed the "performance breeds," and they do pack a punch when it comes to carcass and growth.

Hampshire

The Hampshire is a British breed with erect ears and that eye-catching white belt around a black body. That color pattern can be traced back to a fad that swept Europe long ago, which saw all sorts of livestock bred with the belted color pattern. It can still be seen on cattle, rabbits, and two breeds of hog.

I was raised in an area noted for Hamp breeders. Within a few miles of the house, literally hundreds of head of Hamps would be sold at auction each spring and fall.

In fact, the first big date I took my wife on was to a Hamp breeder's auction. The show started with the sale of a single young boar for $15,500. I recall that with a bit of surprise and awe, Phyllis turned to me and asked, "They can die, can't they?" An old gentleman sitting behind us patted her on the shoulder and said, "They can and they do, young lady."

With the Hampshire breed, litter size has sometimes been a concern; boars were once used far more often than gilts by the commercial sector. We owned a Hamp female of exceptional merit as a mother, and we found that at birth, Hamp pigs are among the most distinctive as individuals.

Hamp boars are regularly employed to boost carcass yield and growth and have been used extensively to formulate show pigs. The Hamp is another of those breeds once classified a "bacon hog," and their trim nature continues.

HAMPSHIRE SOW

Berkshire

The Berkshire is a Black breed with white points on the head, feet, and along the bottom line. Selective breeding has given this old breed a bit of a rebuild over the years. Long gone is the sharply pushed-up nose once associated with this breed. The amount of white coloring has increased as well.

The Berkshire is a breed long known for length, carcass quality, and trimness. A bit smaller framed than some, it does enjoy a reputation for fairly good mothering. This is another British import; a review of the historical literature makes reference to a Red Berkshire bred in the upper South.

The Berk produces a very desirable pork product with exceptional eating qualities. Berk numbers have tended to ebb and flow over the years, and along with the Black Poland, formed the two dominant U.S. breeds prior to World War II.

Berkshire sows performed quite well back in the day, producing their fair share of ton litters, even when the main ration ingredients were open-pollinated (OP) corn and skimmed milk.

Berks can fulfill much the same role of the Hampshire boar in a breeding rotation and can add remarkable type to show pigs. When fitted, Berks present with a real style of their own and have a whole lot of eye appeal.

Spotted

The Spotted is just that: a black spotted white hog with drooping ears. These have always been a large outline breed and are considered by many to be the best mothers of all of the colored hogs. They are the ideal breed for those who want variety in their lives, as no two animals are ever patterned the same.

Our Spots were always good natured and produced that old-style, sandy red hide with black splotches when bred to a Duroc boar. This is a breed for which it is important to select for a wide-floored body and to make sure the back legs are out on the corners. To many, the Spot is the classic farmer's hog, with an extensive history of outdoor production.

The British ancestor of the Spot is the Gloucester Old Spot, which is called the "orchard hog" and has a long history on the small holdings of England and in the simple facilities there. It is no surprise, then, that this is another breed that has been taken up by the British royal family. A few of these animals can be found in the United States, however, and their pedigree has been recorded with the Spotted breed group. Images of Spots frequently grace various livestock literature such as breed preservation brochures and advertisements.

Our Spot sows were good mothers and produced decent-sized litters of distinctly individual pigs. Some say that the Spot was the hog that never crossed the big river. Indeed, it seems to be a genetic resource that is more valued and used in the eastern side of the Corn Belt. The animal brings good length and dimensions; some of the best ham structure I have ever seen was on Spot boars. They are no longer bred in great numbers, but they are a strong Black contributor that should be given serious consideration by the range producer.

Black Poland China

The Black Poland China has been on a very long slide since its high point in the 1920s, when it may have been the most widely bred and highly valued of all pure swine breeds. They were and still just might be the top contender for the title of best all-around farm hog.

Black Polands are durability personified; at one time, the Black Poland herd book documented more sows at eight or more parturitions still active and producing than any other herd book. The Black Polands are black hogs with white points and modest-sized, drooping ears.

The Black Polands I have seen always seemed to epitomize balance, while holding to a fairly lean type. Some needed to be made wider and a bit bigger all over, but they were a pretty complete package. Some really impressive Poland barrows have been driven out over the years, and they do have real eye appeal.

Numbers are way down on this breed, and through no real fault of their own. This is a breed that really needs some energetic folks to take it up and put it back to work. The range producer has to appreciate that the durability quotient and seedstock will progress in the hands of those producers who have clung to a more traditional approach to swine production.

Atypical Breeds

Ag texts from the '30s and early '40s list a number of swine breeds that have closely flirted with extinction, if not actually crossed forever into that vale. We can only speculate what might be done now with a breed like the blue-gray Sapphire. What a marketing tool the color alone would be. The Ohio Improved Chester disappeared while I was still in public school, the remnants of that breed being absorbed into the Chester White breed. Some of these breeds could perhaps be bred up again, or remnants may yet be found in some odd corner of rural America. The quest for rare poultry breeds has shown that this can still happen.

Mulefoot

The return from the brink for the Mulefoot hog shows just what can be done (I kind of had a ringside seat for this one). Mulefoot crosses passed through Missouri sale barns fairly often into the '60s, some with two and some with four closed hooves. Most producers chalked them up to just one more thing you had to contend with when trading in rough, Southern hogs. When corn got cheap, ridge-runners would eat it up just like good hogs.

Gradually, even those animals disappeared, but the legend of Mulefoot hogs hung on in the Missouri vernacular, in part because one of the last three Mulefoot breeders to remain was a resident of a town called Louisiana, Missouri, the riverine hometown of the fictional Finn family. (The old gentlemen continues alive today, but asks that his name not appear in print.)

This farmer's efforts to preserve this venerable breed, one of the oldest of the swine purebreds, came to light about a decade ago, with the early stirrings of the livestock breed preservation movement. Each year, the farmer raised a few as market hogs, with an end-game plan of sending his herd to slaughter upon his death. Also in his possession were some of the breed's historical documents. His line of hogs continued the breed's characteristic black with white points, drooping ears, and four feet crowned with hooves like those of tiny ponies.

One bit of lore holds that those solid hooves were valued because no disease could enter there, as might happen with the cloven hooves of other hog breeds. His hogs were raised out-of-doors in all the various elements that make up the Show-Me State's nearly legendary weather.

Many other breeders have bought hogs from him since then, and these animals can now be found in several small herds scattered about. They have undergone some selection for type, and while none of these will contend in a major barrow show anytime soon, they are again producing some Number 1 butchers. The old literature sometimes mentions red and spotted animals, and a few folks report producing the occasional red-tinged animal. Thus, nothing is ever completely lost, even if just a few viable remnants remain.

Razorbacks?

Word has come down the line lately that some breeders are now attempting to propagate the razorback and promote it as a bit of porcine Americana. I certainly wouldn't want to be in the show ring for a drive of senior Razorback boars. I'm not really sure that the case could be made for a fully Razorback breed, since feral hogs have had the run of the American outback since Colonial times, picking up dashes of new blood with every escaped pig or abandonment domestic hog.

A few months back, I met a fellow who had stumbled upon a very unique niche market for hogs of an uncommon breeding. He had found a market for the young males from litters out of Tamworth females bred to Russian Boars—they were being sold to shooting preserves for fee hunts.

The revival of interest in outdoor swine production has come at a fortunate time, as the variety of pure breeds available still remains large, and animals of desirable type can be found in all breeds. The range producer will have in the purebred boar a genetic resource that is easily introducible into the existing herd and the potential to effect fairly rapid and dramatic change in animal type. With a second, small herd, breeders can even meet their own herd replacement needs.

Good pork production should be about the animal, and range production is all about the animal. Whereas the confinement producer is cranking out a mere commodity—one that is badly devalued and underappreciated by nearly all who are exposed to it—the range producer is a raiser of hogs, one who sees the animal and all that leads up to and proceeds from his or her point of involvement.

This business of hogs and people has gone on in one form or another for about 6,000 years, give or take a century or two. We have come to a point where the consuming public has a collective voice and the disposable income to make some very forceful statements as to how and what they want to be fed.

They want pork from a family farm, and not some sheet-metal gulag with ties to the corporate world. Thus, the independent farmer must strike out on a course that will lead back to the clearly expressed desires of the consuming public. The methods, resources, and genetics to do just that remain in place for at least a time.

This may very well be the last time we get the chance to do it right, to do it for all time.

CHAPTER 4

HERD MAINTENANCE

The smart range producer will draw from the breeds noted here when creating a swine operation that is designed to produce more than mere commodity, but rather, to raise healthy animals that are in keeping with the faith and trust the consumers of America long ago put in the farmers of this nation.

To truly succeed, American pork producers must live up to the trust that has been placed in them, a trust that is implicit in those ham sandwiches packed into school lunch boxes and in the Christmas hams that embody the fullness and assurance of the American lifestyle. Range production and its modest numbers are thus not excuses for poor performance. It is the answer to what is now a very clear mandate from the consuming public.

The range hog must live, grow, reproduce, and thrive to its natural potential in the open air and with simple facilities. It must be bred to the challenge and then undergirded with the kind of humane and thoughtful care that is in keeping with that old Biblical tenet of responsible stewardship.

The circle of life has always been clearly embodied in production agriculture. There is no beginning or end to it. Most folks now farming stepped on board while the previous generation of farmers was still in place, and we are doing what we can to get the next group on board and in sync now. We will never totally be done with the task of farming.

The animals can always be bred better, management can always be tightened up, and the challenge to generate not maximum but optimal production is never ending. I am not trying to make hog production sound mystic or Zen-like, but it really is the essentials of life as a business. The business end must be tempered by nature and humanity, and the system must also produce income sufficient to reward the producer's good efforts. The ledger must tell you where you stand and then must be combined with your head and heart to tell you where you are going.

To this end, you must have a plan, two plans actually. The first is a plan of purpose, and the second a plan of action. With all that is in you, you must be able to answer loudly and clearly every morning you are on the land, why you have the animals around you and what you intend to do with them. And once you can do this, you'd better make sure you can keep track of 'em.

Animal ID

There is a very real need to be able to quickly identify every animal in the breeding herd and to clearly denote the different groups of growing and finishing animals. Lose track of the due date for a bred sow and you're in for long days of uncertainty and fretfulness. An entire litter may even be lost if it is dropped in the sow lot or pasture instead of the farrowing quarters. Or try to commit to memory for even a few weeks which of four littermate Duroc gilts farrowed the small litter and needs to be culled.

As a first choice for individual animal identification, ear tags appear as a good option, and those big numbers certainly appeal to those of us who are now a part of the bifocal generation. These have their drawbacks, however: they aren't cheap, the animals have to be restrained to attach them, and they are rather easily lost. Our county fair ear-tags the market hog show pigs each year at a late-March weigh-in, and by show time in early July, as many as 10% of the young hogs may have lost their tags. And these are animals often penned separately or in groups of just two or three.

The tags can tear loose upon becoming hung in fencing or feeding equipment, or they can be pulled out by other hogs in the pen. Ears are also often split or otherwise injured when this happens.

There are small, quite durable tags that pierce and clamp securely to the edge of the ear. But these, too, are not without their faults. They are very small, can be blurred with dirt, and often can't be read unless the animal is confined in a crate or behind a gate. These types of tags are quite commonly used to trace animals given health treatments or inspections.

Another style of tag that has proved to be a more practical choice is the round, button type that is affixed in the center of the ear. They are a bit over an inch across and can display up to three digits. A yellow background with black numbers gives perhaps the greatest readability of all tag choices.

One of the oldest methods of swine identification, ear notching, still remains one of the best and most lasting. When done cleanly, they are fairly readable from a short distance. Considerable information can be encrypted into an ear-notching system, and they are a requirement for breed association registration. Early ear notches were used to denote ownership when most livestock was still allowed to range about freely.

Early ear notching was done with a pocketknife, and I have done that myself when one or two notches for simple identification was needed. The best and easiest-to-read notches are made with a V-notcher, which closely resembles a paper punch. Good ones can be bought for $20 or so. Make the notches square with the edge of the ear and work the notcher crisply.

Ear notching can be done when the pigs are but a few hours old, when they are being handled for the first time for teeth clipping and iron shots; it is a relatively painless process. Be sure to wipe down and disinfected the instruments after each use.

The notches along the various edges of the ear have different numerical values. Normally, the notches in the right ear indicate the number of the litter into which the pig was born. The notches in the left ear designate each pig within that litter. Poorly made notches may close back, or they can be lost to ear tearing and some other ear injuries. Ear marks have even been used to track the animals through the slaughtering process, up to the point at which the head is removed from the carcass.

Ear notches can be coupled with ear tags, or further information can be included via punches of different patterns made into the ear cartilage. Ear notching is one of those skills every range producer needs to master and then put into regular practice.

There are a number of paints, sprays, and markers that can be used to put temporary identification marks on hogs. Some will last several weeks, and others begin eroding away in minutes. Yellow and green paints and simple, press-on paint brands are often used to keep track of the individual hogs entered into exhibitions and sales. Crayon sticks are useful in keeping track of hogs in a group as they are worked for various types of health measures. A simple swipe on the back can keep you from giving the same hog multiple injections.

Green and yellow seem to be most visible marking colors when used on hogs. Use green paint or crayon on white hogs, and yellow on colored animals. There are a number of spray markers now available, too. Paints in pump-type bottles may work best, as the hiss of an aerosol spray can give some animals a start.

Herd Records

With each animal in the breeding herd easily identifiable in the field, you are then able to start taking down performance data and notes regarding their behavior. Using a simple shirt-pocket notebook, you can log field notes and observations and then enter them into your permanent records at the end of the day.

At the breeding pen, you can note the date when the boar was introduced, exact breeding dates if observed, ID of service sire, ration changes, health treatments, and general observations. At the farrowing quarters, various birth dates, litter size, numbers of boar and gilt pigs, birth weights, feed changes, health treatments, and more should be dutifully noted. The work of a few moments will prove invaluable in evaluating matings and breeding animals, future herd replacement selection, and culling.

At one time, we had a herd of 25 Duroc females that were virtually all full and half-sibs, and it was impossible to keep them separate as we went around morning and evening chores. We filled a lot of little notebooks in those years, but they all fit away neatly in those little file boxes used for index cards

Experience has made me a real stickler about keeping good breeding herd records that are easily usable. With even a very few head, there is just too much to chance to memory. And the best way to fight a case of barn blindness is with good, solid figures in hand.

We hear a lot of razzle-dazzle now about performance indexes and herd averaging, but often these numbers are based on some sort of statistical straw-man rather than a real-world norm. Indexes usually measure variance in performance based on a group norm. If that group is small, managed specially, or is otherwise skewed, however, the index numbers are of very little practical value. They will lack true validity until they represent data collected from generations of animals.

The old-timers that I respected and learned the most from based their breeding herd records on some very solid, economically important norms. At birth, they weighed individual pigs, ear-notched them, checked underlines for teat count, and made their first sort for those that they believed would be the keepers. A pig 3 pounds or heavier at birth had a far greater chance of survival, and a pig much under 2½ pounds was fairly quickly eliminated from any sort of future consideration. Uniformity of the litter is always a crucial factor on the small family farm.

In the days and weeks following the birth, herd records would include losses during lactation, sow condition, pig growth, and health care practices as they were delivered. They would log in pertinent weaning data such as litter size and weaning weights. Days of age to slaughter weight, ration notes, feed efficiency data, and the like would then follow. It is fairly simple to keep track of the pounds of feed delivered to a finishing pen or taken out of the sack to feed little pigs.

For literally generations, farmers successfully managed swine operations with data easily collected on the farm and with simple, easily replicated methods. As selection technology supposedly improved, it became more costly, stirred up a lot of controversy as each new method drove out a few more producers, and pork quality often rolled and yawed badly in the process. Now, even the university boar-testing stations have all but passed from the scene, and considerable indexing is now being manufactured.

My skepticism about so-called performance data began a long time ago. I was still in high school and attended a Duroc breeder's presale program where MU staffers were demonstrating back-fat probing as a selection tool. At that time, it required a special holding crate, the handling of near-market-weight animals, and incisions and various probings at three different points along the animal's top-line. At the time and down to this very day, most farmers are reluctant to subject animals to so much handling stress and the drawing of blood.

In the question period that followed the demonstration, an old gentleman sitting high up in the bleachers raised his hand. He said that due to age and personal concerns, he was probing his gilts only at the back position and added one-tenth of an inch as a selection guide. This he decided upon after a long study of his own records, it wasn't done on any animals he offered for sale, and was used only to evaluate animals returning to his own herd.

The university staffer jumped on that old gentleman like a duck on a June bug. He was told in no uncertain terms that it was a practice that simply could not work in any way, shape, or fashion. But only a few months later, I noticed a press release issued with much fanfare in which university research had determined that adding one-tenth of an inch to the back fat (or "bf") reading at the third position would be a very good gilt selection practice.

Breeding Selection

On a fine June day several years ago, I stood in the crowd at a Duroc picnic and watched two of the big players in the Midwestern swine scene come to near blows. The highlight of that day was a swine judging contest, and one of its features was a very strong early-spring gilt class. One gilt in the class was exceptionally long and lean for that breed and that day and time. She went to the top of the class on nearly every judging card turned in for scoring.

Official judges included a prominent breeder and a field rep for a major packer. As they stood together to officially rank the class, their discussion became ever louder and more animated. Finally, the breeder was told to explain the class if he continued to insist that it be placed his way.

When he placed that good gilt last, you could have heard a pin drop. When he gave his reasoning, saying that she was far too extreme to be even remotely practical, it got quieter still. The following fall, when he paid big dollars for a gilt described in the publications of the day as being one of the most extreme representatives of her breed ever seen, I got quiet again and then had a good laugh.

At another breeder's sale, we sat through a once-common practice of having the herd boars driven out to be oohed and aahed over. Among the boars was one that was determined to be the highest-testing boar to date produced in the Midwest. To be honest (and generous), he sure wasn't pretty, and there were some fairly obvious soundness issues with him.

The two old guys sitting behind me summed him up pretty well when one said to the other, "Don't you think they should've spent more time looking at the boar and less time looking at his numbers in the catalog?"

Numbers do not tell everything there is to tell about a hog, good or bad.

I have never seen any part of a farming operation go completely by the numbers, and there is no index or formula to plot that most important economic trait of all, live young. And of course there are some selection factors that can only be determined by visual appraisal, and the word from the breeder has to be relied upon when trying to determine what the animals up for sale can and cannot do.

When we were producing purebred boars in good numbers, we found a niche market that the performance breeders had created quite inadvertently. There was and still continues to be a market for what can be simply termed "tough" hogs. It was an unspoken truth that at the height of the performance-testing mania, producers kept a replacement boar at home for every one they sent to the test station. Those severe, numbers-driven environments generated dazzling figures, plus a great many dead-headed, dud boars. They regularly had to be replaced with boars produced in a far less stark and driven manner.

When I took a buyer to see our boars, I pointed out that they weren't fitted, they'd been fed a pretty standard grind-and-mix ration, wouldn't win a lot of shows, they knew what it felt like to walk on dirt, and in 30 plus years, we had replaced fewer boars than you could count on one hand. They were as tough as I could make them.

To add a new Duroc breeding line a few years back, we went through a good half-dozen gilts before we found the one that put it all together for our little herd. We had some pretty good Spot sows 20 years ago at the height of the flat muscling craze, and we never did find a boar we felt comfortable about buying to mate with them. You have to be sober-minded when taking the measure of animals from which you expect to make your livelihood, and yet remain open to important genetic opportunities when they occur.

Once on a trip shopping for a boar, I violated a personal rule and bought one that had been shown nationally. I was on a farm with a long-established Chester breed that had a herd that was decades old. We were only about 90 minutes from home, and they were thus from a similar environment and system of production. We had also worked through several groups of young boars that day when he suggested I view a boar that they had just brought back from the National Barrow Show.

All kinds of personal radars went off, including the nagging question of why he hadn't been sold in the auction there. The boar had placed fourth or fifth in his class and thus was pretty well out of the money for the sale of that particular breed. He was a big-framed old pig with a lot of length, but perhaps a bit thick for the times. I figured that if I could get him home and on a breeding ration, he would trim up nicely. He represented the animal type our regular buyers were looking for, and he fit our boar-buying budget. That time, I broke my own rules, and it worked. The many times I did it and it didn't work, however, would take far too many pages to cover here.

We have also owned a few that were as ugly as homemade soap, but when they were bred right, they went on to produce some exceptional offspring. I sometimes felt that Dad went out of his way to find those kinds. We once had a real rat of a runt pig that he managed to run between my legs every time we had a high-browed visitor to the old place. It did go on to serve as a good reminder that a lot of hogs offered for sale are presented in a highly fitted and conditioned state. A hog in its working clothes will sometimes look very different than when first glimpsed in a sale pen.

I was once fortunate to see a three-generation display of one breeder's efforts to create a line of Hampshire breeding hogs. The presentation began with a drove of 60 pound shoats of wonderful type that were as uniform and shiny as a roll of new dimes. When sire and dam were driven out, you could see strong type similarities to the shoats. The sow was very milky in type and was barrel bodied. The boar was long, with a well-defined muscling pattern. You could see some of the bits and pieces they had passed on to their offspring.

When the grandparents were driven out, we were all in for an eye-opening experience. One boar was very extreme in length and appeared to be big-structured, almost all bone. The other was a textbook example of breed type and character. One sow was quite the rugged old rip and was very loose sided. Pretty is as pretty does, though—she had a 10-plus weaning average. The other was still showing exceptional muscle definition and trimness, despite her age. It was one of the best "big pictures" of what well-planned swine breeding should be that I have ever seen.

Don't hesitate to cull ruthlessly or perhaps feel bad about building extensively on a single animal if it's one that suits you. It is not uncommon to see a good herd line that traces back time after time to a single foundation animal. Generally, it is a female; we have owned sows that produced good pigs when bred to several different boars.

We have never followed a true line-breeding program, in that we have always tried to fulfill the role or niche of a multiplier breeder. We sought out animals of practical type from lines and pedigrees that had proven worthy over time. Real-world family farmers learned long ago not to be impressed by purple ribbons and magazine covers. We still have breeds today like the Mulefoot and Tamworth not because they were "front pasture" animals, but because they were tough old hogs.

Swine breeding truly is a never-ending pursuit of a better specimen. The perfect animal hasn't been nor ever will be bred. If it were, that would take all of the fun out of it.

It can literally take years to measure the full impact of a single animal on a breeding program. We have waited to cull boars until after their first pig crops reached slaughter weight, and others were never even allowed to reach the breeding pen. We will give a gilt a second chance if we believe her performance was adversely influenced by environmental factors, but if these animals fail the second time, they have to go.

The Breeding Lot

The two great tests for any breeding herd are in the breeding pen and the farrowing quarters.

A gilt will start cycling at about 6 months of age, but breeding should be held off until the gilt weighs 300 pounds at around 8 months of age. She will farrow about 4 months later, at 1 year old.

Some outdoor producers hold off even a bit longer and will not breed the gilts until they are 9 to $9^1/_2$ months old. That extra time converts to added body development and extra size. The extra size gives the still developing gilt a bit of a reserve to draw upon during gestation and lactation and should help her to breed back more quickly following weaning.

Some farmers have had real problems getting second-litter gilts to breed back quickly, thus knocking herd schedules for a loop. For best results, you really have to keep your females well bunched and moving through the year and the facilities as groups. The old rule of thumb has been to feed a bred gilt enough to assure a weight gain of at least 125 pounds during her 114 day gestation. In subsequent gestations, you want the females to gain at least 75 pounds to fully recover from being nursed down. Bear in mind that those second-litter gilts are going to need a bit of extra attention, as they have not yet truly graduated to sow status.

Sows in a group will establish a pecking order much like chickens, and the range producer must always be mindful of it. Newcomers to an established group can set off any number of problems, from a few days of simple stress to vicious fighting and major injuries. For this reason, when natural attrition begins to reduce numbers in a group, many producers will opt to replace the entire drove rather than go through the hassle of incorporating replacements.

Still, sometimes it's easier to work in new animals. A few tricks can help ease the transition.

1. Begin with an extended period of contact through the fence.
2. Introduce them only when you have the time to supervise the blending and watch them for an extended period of time.
3. The females should be as close to each other in size as possible.
4. Do it in as large an area as possible, and work to provide outs and escape points to which animals can retreat.
5. Some farmers will lightly spray all of the females with kerosene or other strong-smelling spray to give them all the same scent.
6. Try to make removals and insertions at the same time to take advantage of the new breakdown in the pecking order.
7. Some producers will wait until the boar is present in a group before adding new females.
8. Try not to insert females of distinctly different colors.
9. Do not mix animals during times of temperature extremes or when footing is poor.
10. Get any animals that are in trouble out of the group quickly and accept that some additions may not be possible.

I have seen sows in groups as large as 25 to 50 head. On most farms, they are kept in groups large enough to match the slots available in the farrowing quarters. To keep things fully on schedule, the big operations breed 10% more females than they have slots for, to make sure they always have everything full and producing to capacity. What they do when they have 30 stalls to fill and 33 sows ready to have pigs is never made clear.

The old rule of thumb, and one I always adhered to as a seller of boars, is that you need to maintain one breeding boar for every 10 sows on the farm. You can cheat on this a bit, and if you have two groups of six or eight sows that are managed to farrow some months apart, one boar may be enough. In pen groups of much over 15 sows, two boars working together may be needed to keep their farrowing dates tightly bunched. Young boars especially have a tendency to fall in love with one female and neglect others in the group.

In those instances where you will need multiple boars to work together, buy boars that were raised together. I have broken up a boar fight or two in my life, and it is not a fun thing to do. It will give your heart medicine a real good test. If the boars are full or half-sibs, you should still get good type consistency in the pigs.

An alternative for large sow groups is to keep boars penned separately and alternate their visits to the breeding pasture every couple of days. Boars stay fresher and you are less likely to have the problem of a boar falling in love and following a single female for her entire heat period.

A young boar can begin service at 8 months of age or even a bit younger if well grown for his age. With their weight kept reasonably in check, they can be kept fertile and active for many years.

Good breeding-boar management calls for an eye to a lot of fine details, but it's worth being careful, since the boar is a full half of the herd and can represent quite a substantial investment in time and funds. Some further notes on boar management are therefore warranted.

1. Don't overmatch a young boar. If larger, more aggressive sows turn on him, he may never breed.

2. For the first services, try to match him to sows of some experience but close to him in size.

3. Observe early services as much as possible. We have owned some boars that are true night breeders, and this seems more common to some breeds than others.

4. Watch closely for penile injuries, signs of blood in the semen (caused by high riding and or abrasion injuries), pooling of fluids in the penile sheath, and flagging libido.

5. Never "ring" a boar (place a ring in the nose to prevent rooting), even one on pasture. He uses his nose to position the female for breeding, and ringing may even void any warranty placed on the boar guaranteeing his reproductive performance.

6. In the breeding pen, feed the boar exactly the same as the sows. Away from the breeding pasture, hold him on a simple maintenance

ration. Without such a diet restriction, a boar can soon grow too large to breed safely.

7. If viewed infrequently, get the boar up and moving about every time you do chores, to better monitor his condition.

Pasture breeding, especially with multiple boars, will create some uncertainties as to both parentage and due dates. Boars going into the breeding lot should quickly trigger the onset of estrus in the females. You should provide sufficient boar numbers to get all of the sows bred in a fairly tight 10 to 14 day time span. Most breeders will then pull the boars after the sows pass through the 21 day period between cycles and demonstrate that they are safely settled.

Signs of breeding activity can include roughened hair and mud on the sow's backs. A semen plug may also be visible in the vulva.

Gestation

Pasture or large wooded lots are the most natural places on Earth for gestating sows. There they get much-needed exercise and the stimulus of a varied and changing environment. On a good legume pasture, they may me able to forage as much as 10% to 15% of their total nutritional needs. In warm weather, you can really back down on the feeding of concentrates and protein to sows that are on good legume pasturage.

You must always feed to the needs of the gestating sows, however. Sows rebounding from weaning big litters will need extra concentrates to return to condition and successfully carry the next big litter to term. The old bromide holds that a gestating sow will need at least six-tenths of a pound of crude protein daily to maintain a good performance curve.

This can be achieved by feeding 4 pounds daily of a grind-and-mix ration (yellow corn and soybean oil meal-based protein supplement) with a crude protein content of 15%. Another option would be to offer the sows 4 pounds of shelled corn and 3/4 to 1 pound of a protein supplement daily.

It is rather hard to ascertain just how much nutritional benefit a sow is deriving from pasture, and I favor offering the sow that same type and amount of ration that would be offered to her in a drylot. The pasture then becomes a nutritional plus. The producer on pasture has the task of matching browse quality to standard feedstuffs to keep the animals in good flesh.

There are times when protein supplements or concentrates can be measurably reduced. In good weather, you may need to feed just ½ pound of hog concentrate per head per day. Corn may be trimmed by a pound or two. And there are times when you will need to feed them just as if they were in a drylot. This is one aspect of range management that cannot be taught; a practice for which there are no formulas and few rules of thumb. You can only be guided by animal appearance and behavior and the weather.

The nursing sow will rob heavily from her own body if she must, and while some thinning is to be expected, she must always be maintained in a weight-gaining condition.

Essentially, the animal retains the health and body weight that will enable her to return quickly to a good weight for her age and state of production. Pasture, good pasture, keeps sows in that potential weight-gaining state better than perhaps any other environment. You must never assume, however, that just because you've put them on grass you can just dump a little corn over the fence and forget about them. They need to be studied often for body condition, pregnancy rebound, development, and steady growth.

To really step up onto the soap box for a moment, let me say this: animal nutrition is not a finished science. We don't know all there is to know about the nutritional needs of the hog or any other livestock species. A number of years ago, a major Midwestern feed company began to supplement its swine feed with a very diverse mix of protein and other nutritional sources. Included in the ration were a number of fish- and plant-based products that added what that company called UGFs: unidentified growth factors.

Vitamins are a quite recent discovery, and it stands to reason that the effort to simplify swine rations for handling and feeding in confinement may have resulted in rations that are not as complete as they should be. We know that hogs will eat grubs, acorns, hickory nuts, woody forbs, and the odd lizard from time to time, presumably as dietary supplements. I am sure, though, that you'd be hard-put to find nutritional data on that odd ration mix.

On range, hogs are getting maximal sunshine, they are fairly free to select from a variety of different food items, and the good producer will be carefully undergirding them with the staples of a traditional ration. The animal is thus able to forage to its own satisfaction and, quite likely, for its own nutritional satisfaction as well.

Even many producers with extensive investments in confinement facilities believe that herd performance is greatly improved by getting the sows out-of-doors at least a portion of each year. And yet, as important as she is to the overall success of any swine venture, the gestating sow still too often gets short shrift.

In a drylot, a gestating sow will need at least 150 square feet of pen space, and in wet climates, this space should be tripled. With Missouri's potential for wet springs and falls, and summers, and winters, I like a figure of 600 square feet or so. I haven't come across any really good rules for stocking rates on pasture.

One key task that's largely left up to the producer's good judgment is to match sows to pasture and to prevent any damage from overstocking. Bear in mind that hoof wear is generally always the culprit in mud problems and lot failure. Frequently traveled paths can soon become ditches, and yet one of the reasons for penning the animals outside is to encourage movement and exercise.

While older hogs don't necessarily get lazy, they are a bit like older hog raisers, in that they slow down a bit except when it comes to getting to the table. To make sure that they are getting out and about enough, the breeding herd should be fed at least 150 to 450 feet away from the sleeping sheds.

The real plus of pasture production may be every bit as much a physiological boost as it is nutritional. In a pasture or lot-breeding situation, the animals will be more content, they will be in better physical condition, will breed and settle better, and they will benefit from a number of near-intangibles. Most sows in confinement are gone from the farm before their fourth farrowing, and I have to believe this is largely because of all the things they're lacking in the confinement housing environment.

In the out-of-doors, the onset of estrus seems to arrive sooner and is more pronounced. The boars are more responsive and fertile. There the animal will encounter a great many stimuli and then respond in a strong, natural manner.

I'm old-school and believe that all a breeding boar should need is for you to open the gate to the sow lot and then get out of the way. Working a boar in confinement has become a combination of art, science, and a bit of black magic. A growing number of confinement producers are relying upon artificial insemination, and still others rely upon breeding barn specialists who use everything from heated up rations to special stimulating aerosol sprays to get animals to complete what should be the most basic of species survival practices. Any boar that needs outside stimulation to breed probably shouldn't be allowed to.

Kept out-of-doors, breeding hogs will fare well in nearly all weather conditions, particularly if they're allowed to become acclimated. We've had sows and boars in open-fronted, cold housing at temperatures 10 to 15 degrees below zero, with no adverse effects. Actually, a cold drizzle with air temperatures in the 40s is more trying than far colder but dryer conditions.

Straw bedding and protection against chilling drafts will go far to keep the animals comfortable and help them better contain their body heat. When out-of-doors, year-round, the animals quickly become acclimatized to the changing seasons. Pull hogs out of confinement in raw weather and they simply pile up like cordwood, and all sorts of problems result.

Midwestern farm lore is full of stories of hogs that escaped captivity and wintered in beds of corn stalks or fallen leaves. They will need dry bedding at a depth that they can burrow into a bit, and added grain to help them generate some extra warmth from within is also a good idea. In winter, try to give them a good feeding and watering late in the day to give them added body fuel for the long, cold night ahead.

Drinking water is a crucial feedstuff and should always be considered as such. Granted, hogs will drink what looks like bad chocolate milk out of cow tracks, but the cleaner and fresher the water, the more they will drink. And the more they drink, the better they will perform. A gestating sow will drink at least 1 to 2 gallons of water per day in cold weather, and 5 or more in summer. A nursing sow may drink far more than that.

There is definitely a work element to sows on pasture. You will have to go out where they are living and invest time in simply monitoring the conditions and their well being. You are going to get rained on, you will live much of your life in knee-boots, and you won't get to go around in those nifty blue short-sleeved jumpsuits that the folks wear in confinement units. You do get to work with healthy and contented animals, however.

The breeding herd is the heart of the operation. It is the genetic storehouse, the pig producer, and the pig-rearer. For this reason, it has to be managed in a very exacting manner.

Once upon a time, the range producer had to keep his or her animals in an almost constant state of motion. Lots were turned often to help control parasites. Modern parasite control methods now reduce the need to rotate swine pastures as often.

When it comes to treatment, I am neither a detractor nor an outspoken advocate for the organic or holistic approach to swine production. I prefer to use some of the products now available to us, but with a certain amount of restraint. We've found that diversifying treatments helps; for example, diatomaceous earth improves the performance we get with some of the wormers mentioned below. But we do take advantage of the fact that medicines have evolved far beyond a dependence on a single product that was largely effective only on one worm species, the large round worm.

New products now make it possible to work with near-permanent dry and wooded lots and pastures. Sows need to be wormed two or three times a year, and the best times are generally a couple of weeks before farrowing. You should also rotate wormer products after one or two uses, to prevent any type of resistance from developing.

Common worming treatments include Ivomec, Tramisol, Piperazine, among others. Some of these are injectable, while others can be added to the drinking water or to the feed. When treating sows, I especially like the feed-applied wormers, but you have to be sure that each sow gets the exact amount she'll need for her weight.

External parasites can be treated with Ivomec, sprays, or dust-type products. You can generally use spray products on breeding stock anytime air temperatures go above 40 degrees, for even a few hours. Dust products can be used on the hogs and bedding in cold weather, when lice problems can be especially bad because the animals sleep so close together.

The producer who has modest numbers of animals and hence no problems with crowding and overstocked lots and facilities will have far less risk of parasitic outbreaks.

Outdoor Farrowing

First of all, "outdoor farrowing" isn't outdoor farrowing. It is when something goes awry and range producers have to earn their spurs—or knee-boots, to be more accurate.

Farrowing is not an overly difficult or trying time, but it's the one time that requires the fullest possible focus on the part of the producer. As one of my friends once put it, standing in a stiff wind with a cold rain running down your collar while you tend a litter of pigs cuddled warm and dry in a range hut tests just how serious you are about being a "hog man."

On pasture, you'll want to set those houses 100 to 150 feet apart and get the sows out to them early enough for each animal to claim one as her own, generally about 2 weeks ahead of their due date. Set those houses with their primary doors facing south, away from prevailing winds.

Even in warm weather, provide at least a modicum of bedding for pig comfort and to keep things dry. As long as the ambient temperature regularly stays above 60 degrees, you can farrow on range without using supplemental heat. Thus, you should be able to keep your operation out there from March through November in some parts of the country, and for at least 6 months out of the year in most areas. And as noted above, there are a variety of supplemental heat sources now available that make year-round range farrowing possible.

The sow on range will not be nearly as prone to some of the ailments commonly associated with farrowing in confinement, such as constipation. Just be sure that she receives plenty of fresh water, and if she has to be kept close for a few days following farrowing because of inclement weather, it's a good idea to add some Epsom salts to the water first offered after farrowing, or add some bran or other bulk to her ration. I'm also a firm believer in the old chestnut that says there is virtually no livestock ration that cannot be made better with the addition of a bit of good alfalfa hay.

As a youngster, I sat up with a lot of farrowing sows and spent many pleasant hours that way with Dad, hearing the stories from earlier days. We saved a few pigs and helped a sow or two, but it wasn't really in keeping with our beliefs and expectations, which was for a more natural system. Like breeding, farrowing is one of those basics of life that the animal should be able to do all on her own.

Many breeders will get overanxious with the farrowing process, and for this reason, the contraction-stimulating drug oxytocin is frequently abused at these times. The drug is also used to stimulate milk flow, but it has a minimal effective life in the animal's system. If a sow is straining with a dead or breach pig, this course of treatment only causes her to tire that much sooner.

We check the farrowing houses first thing every morning and last thing at night. If all seems well, then I just have to trust in our selection process for breeding females. Once the sow settles into the farrowing process, it's simply a matter of letting nature take its course.

Once farrowing has begun, a pig should arrive every 20 minutes or so, and a normal farrowing should be finished in a couple of hours. Small litters will often be carried past the normal 114 day due date, and large litters will come a bit early. Some breeders hold that the small litters will have a greater percentage of boar pigs, and that the reverse is true with large litters. The smallest pigs in the litter generally also arrive early and late in the birth order.

A female that has strained for an hour or more without producing a pig, that has bloody or foul smelling vaginal discharge or is in obvious pain is in trouble. With experience and training, you can learn to assist with some troubling presentations, but this sort of knowledge is best learned shoulder-to-shoulder with a patient veterinarian. I have been talked through a couple of rough deliveries over the telephone, but I've also had a couple of close calls when I waited too long before summoning help.

I honestly feel that many of the health care basics can be mastered by nearly anyone, and skill number one is to quickly assess the situation and your ability to deal with it. If you're going to need help, get it in there as quickly as possible. Yes, a vet call costs money, but it will cost far more if you drag your feet and have to call the vet out after hours to deal with a real train wreck. The quicker the treatment is sought, the sooner recovery can begin, and the costs to treat are often much less.

Litter size is largely a function of environment, and the queen bee of that environment is the sow. Savagery, skittish behavior, or deadheadedness can all take a toll on young pigs. Overlay is a major pig killer; I have seen sows that would simply flop straight down and then be totally unmoved by the squeals of distress from the pigs caught beneath her. On the other hand, we once had a Chester sow that made a point of lying down well away from her pigs and then calling them to her side. This is one of the strengths of range production, in that the animals have the room to follow such natural tendencies.

It takes time to select for and to build a strong mothering line, but it is a must for the range producer. There are no shortcuts to this task, and you may find that you have to turn loose of several females before finding the one or two sows on which you can build a herd.

The hands-on role for the producer begins shortly after the little pigs arrive—things really get busy when the little guys are 12 to 24 hours old. Of course, not every producer will follow the same practices immediately after farrowing, but here are some of the more common early life management methods for baby pigs.

1. The eight needle teeth should be clipped down even with the gum line. Some call these the "wolf teeth," and they are found on both sides of the jaws. Their removal will prevent discomfort to the sow's udder, as well as limit any potential fighting injuries and resulting infection to the little pigs. I favor a small pair of side-cutters that will nip the teeth cleanly, without crushing any of them.

2. Some producers will clip tails, and this may now be a requirement with some marketing programs. I don't recommend tail docking of herd replacements, both for aesthetic reasons and the fact that there is strong evidence of prolapse and other problems later in life.

3. Ear-notch the little pigs within the first 24 hours for minimal stress.

4. Begin logging in individual data with the first handling of the pigs. You can begin individual animal records with the ear notch, birth date, sex, birth weight, teat count, and more. I was able to pick out most of the boar pigs we would retain for sale on the day they were born.

5. Iron injections are sometimes administered in the early days of newborn pigs, typically in the first 72 hours. This may not be necessary, however, for pigs that have access to clean ground. If you do go ahead with this, the injection should be given in the long muscle at the side of the pig's neck. This is to avoid staining problems that have been known to occur with iron injections into the ham. Select an iron type that will flow well and is readily taken up by the little pig's system. In raw or cold weather, a second iron injection may be needed 10 to 14 days later. Also available are a variety of oral iron products, but the young pigs don't seem to take these up as well as most injectables.

6. An old trick for an alternative iron application is to drop a scoop of clean sod into the pig area every couple of days to enable them to get some natural "rootin" iron. At one time, barrels of such sod were gathered each fall for use with winter and early spring arrivals.

7. For market hogs, the sooner the castration procedure is performed, the less stressful it will be on the pigs. I have seen this done quite often in the first week of life, although it will take a bit of care when working with little pigs.

8. With wooden-floored housing and rough pen surfaces, be on guard for navel ailments caused by abrasion from a rough surface or from scrambling over a low barrier. To that end, the navel should be

treated with an astringent shortly after birth. Half a cc of LA 200 in the long neck muscle opposite the iron injection site can also help to prevent this problem.

The tasks carried out from this point until weaning are largely to see to the pigs' comfort and nutritional needs. At somewhere between 10 and 21 days of age, the pigs will begin venturing out from the farrowing unit. Over that time, supplemental heating for the young pigs can be gradually eliminated. This is a process commonly referred to as "hardening off." By 14 to 21 days of age, the pigs should no longer need extra heat.

At this stage, many producers will move sows and pigs out from a central farrowing facility to a range or drylot environment. The pigs will then be well hardened off, and there should be no problems with piling (which is when they climb on each other when scared, crowded, or cold).

Be sure to provide extensive square footage for each sow and litter in group housing, and place no more than two or three sows and litters to a range house. Many farmers will gate off such houses for a few days to make sure the sows and litters get a clear impression of which house is home. It can be heartbreaking to do morning chores and come upon six or eight sows and their litters crowded into one range house, with many pigs lost to piling and overlay. Don't skimp on housing space here.

Use generous amounts of bedding, but not so much as to tangle or bury little pigs. You can add to it later as needed.

If the pig is kept dry, comfortable, and well fed, it is free to concentrate all of its energy on growth. However, the producer needs to be on guard for respiratory and scouring problems and signs of slow growth. With scouring, a pig can quickly dehydrate and die; it is symptomatic of a number of health problems, from simple milk scours to the often deadly TGE (transmissible gastroenteritis).

While working for a local elevator, I gained a bit of a reputation for curing pig scours on a number of Amish farms simply by replacing a pig starter formulated for confined pigs with an old-line product with a lower crude protein content.

Little pigs don't have a lot in the way of reserves for fighting an illness. The real trick to caring for them was pretty well set down by my father when he said, "Treat 'em not just when they're sick, but when they look like they're just thinkin' about it."

Simply put, the producer who's in tune with the animals can sense when they've been exposed to a health threat or have otherwise been stressed. That's the time to step up the pigs' comfort level, get vitamin/electrolyte products into the drinking water, and begin with standard preventative care. Don't hesitate with small pigs, since you can lose a lot of ground in a very short time. A lactobacillus product can also prove helpful in maintaining good gut activity.

One method practiced long ago to treat a scouring pig was to administer charcoal and buttermilk. Not only was it rustic and colorful, but it also had some merit. The charcoal would slow the gut and absorb some fluids, and the buttermilk could help with improving the intestinal pH.

For quite a spell, we had to work around a problem of white scours that hit a few pigs in some litters at about 3 weeks of age. It was noted that this was 3 weeks into the sow's cycle, at which point she would have otherwise entered her estrus cycle. This was probably triggering some subtle chemical differences in the sows' milk and causing some gastric upset in the nursing pigs. We began changing the scheduling for our management practices around this time. We made a point of not moving sows and litters, we kept pig handling to a minimum, and made sure there weren't any ration changes, and this helped a great deal.

Unlike some producers these days, I don't believe in giving shots just for the sake of doing something. There are other things that can be done to help keep pigs moving along. Lactobacillic products are often quite useful in fostering good health, and they come in both dry and paste forms. The paste form is quite useful in dealing with individual animals. There are also some energy supplements and other nutritional boosting products that can be used to help smaller and poorer performing pigs and shoats. I've early weaned a few little guys over the years, to tighten up their environment and give them some special care, but I find bottle feeding awfully hard to justify.

Pig starters and growers now seem to come in more forms and varieties than the penny candy of my youth, and now they even come in strawberry and chocolate flavors. There are pre-starters, starters, starter/growers, and growers. With some of these products, you are meant to feed only a pound or two per head before changing to a different ration. Some are liquid, softened, crumbles, mini-pellets, and plain old pellets. Some are extremely high in protein and quite complex in their formulation. And none of these are cheap.

I am not necessarily anti pig feed, but I am opposed to buying a Cadillac when all you really need is a Chevy. One of the factors I believe that determines starter usage is the age at which the pigs will be weaned. Pigs will start to get curious and mimic the behavior of their mothers by about 10 to 14 days of age. If weaning at 5 weeks, I want to encourage their feed interest a little and will go with a more flavorful and richer starter. With an 8 week weaning, a starter/grower may be the better choice, but you want to make sure they're eating really well to keep them from taking the sow down hard. The old rule of thumb we used was that by the time a pig weighed 40 pounds, he'd have one 50 pound bag of pig feed in him, and that guide still holds up pretty well.

Weaning Time

Weaning is always a time of potential stress in a young pig's life, and it can be the same for the producer, too. For at least a week before weaning and another week after, don't make any changes in rations or otherwise subject the pigs to any other changes beyond the actual removal from the sow.

There is a school of thought that runs contrary to this, which holds that it is better to get everything over and done at once. We've had neighbors that would wean, sort, worm, castrate, and relocate the pigs all at one time. They had few losses, but I always suspected that it hurt the growth curve, postponing the pigs' development at a time they should be doing some of their fastest and most efficient growth.

Some producers, however, take an even different tack, weaning over a period of days. They remove the largest pigs first and give the smaller ones a few extra days of nursing.

Perhaps the ideal way to wean pigs is to remove the sows and leave the pigs in familiar surroundings. The sows should be completely removed from within sight and sound of the freshly weaned pigs, both for your peace of mind and for the sake of your fences.

Some producers will remove the pigs to a separate tightly fenced lot or pasture. Another option is a "cold" pig nursery. This is typically a wooden-floored range house pulled adjacent to a slat-floored pen of the same dimensions. The goal is to provide 4 square feet of dry, mud-free space for each weanling pig.

The key to successful pig weaning is to keep the pigs comfortable and secure. Straw bedding at least 4 inches deep will add considerably to their comfort; it will keep them looking clean and bright and will give them something to nestle and play in. To avoid any risk of gut impaction, some producers will bed with fine hay rather than wheat or oat straw.

The feeding equipment should be a type they are already familiar with. If you're using a feeder with lidded caps, tie a few of the lids open for a time, at least until all of the pigs get the hang of operating the feeder.

The drinking water needs to be fresh and clean, and consider the addition of a vitamin/electrolyte product to help them through these potentially stressful times. Animals will often continue to drink when they won't eat, and this supplement will help them to maintain a good condition. To foster and increase water consumption, you can even add a box or two of flavored gelatin to the drinking water.

A Time for Taking Stock

Weaning is a good time to pause and take stock of how things have performed up to this key point.

At this stage, small and uneven litters are clearly evident, and the memory of how well individual sows did or didn't perform during farrowing and lactation is still fresh. I will give most gilts a second chance after a rough start, but a poorly performing second-litter sow is likely to set a pattern of performance that will probably rob from the bottom line for as long as you own her.

I also believe in taking a hard look at the pigs themselves before moving them on to new quarters or a new owner. No set of pigs is ever completely even developmentally, but your goal should always be to keep your groups as close together as possible and moving through your system apace. Small pigs and late farrowings will hinder a group throughout its time on the farm, since younger and smaller pigs are always at a disadvantage in group situations.

Stragglers can be dropped back to later arriving groups, but there are still apt to be problems there. The best option may be to move the animals to a separate place, grow them to a good feeder pig weight, and then send them on a one-way trip to the local auction barn.

For the feeder pig producer, the time for cashing in arrives shortly after weaning. Decisions must then be made regarding selling weights, how the animals are going to be sorted for quality, and how to treat for external and internal parasites. External appearances can be a significant selling factor in the feeder pig game.

At the moment, the feeder pig trade is in very uncertain straits. It has always been among the most volatile of all livestock markets, but currently it simply lacks any sort of dependable definition.

A great many pigs are now early weaned and locked into contract marketing situations. They move from the farm where they're farrowed, taken to another farm with hot nurseries, and from there they typically go to yet another farm with finishing facilities. There is no posting of prices or reporting of market exchanges with this system.

A few pigs may pass through the remaining mixed-stock auctions, but in nowhere near the numbers seen even a few years ago. And there the markets are truly based on who is in attendance and just how badly they want a few pigs to feed. When early weaned and hot-nursery pigs find their way onto the open market, they are generally quite sharply discounted, due to their delicate nature.

Back in the dark and dirty mid-'90s, a lot of hot-nursery pigs would pass through area sale barns without finding a single bid and were often given away. At that time, a friend of mine bought a set of 100 pounders off of one of these lots—she lost 40% of them in the first night out in a fairly warm Missouri December.

Missouri no longer calls itself the "Feeder Pig Capital of the World," but here and elsewhere a pig market of modest means continues. As I discuss in some detail below, the local industry has been built on a 40 to 50 pound, hardened off and ready-for-the-dirt pig.

Recent research has shown that the pigs raised out-of-doors have strong beneficial traits that hold up throughout their lifetimes. They may consume a bit more feed per pound of gain, but they clearly emerge as hale, robust performers that are tough in the good sense of the word.

In the outdoors, 15 to 20 growing hogs can be stocked per acre, and ideally they'll derive a minimum of 10% of their nutritional needs from good legume pasture. Unlike cattle and other ruminants, hogs can be placed on pastures that are composed of 100% legumes. There is no risk of bloating or other digestive maladies when running them on stands of pure legumes.

Still, to ensure top performance, they should be offered a complete growing/finishing ration to match their appetite. A complete 15% crude protein ration will keep them growing rapidly up to a good market weight.

Days to Market Weight

The number of days to market weight are also the days to the next paycheck, and thus are very important to the swine producer. Indeed, much has been made of these numbers over the years, and there is a lot that can be done by the producer to shorten a growing hog's time on the farm.

The issue is just how to pare those days to a market weight in a truly cost-effective manner. Growth rate has a fairly high degree of genetic inheritability, and thus it is a trait that responds well to selective breeding. It can also be boosted by nutrition and environment.

Complex, high-protein rations (17% to 18% crude protein; lots of dairy product) will push growth rates, but in the latter stages of development, when growth naturally slows, those pounds will go on at a heavy price. Very lean hogs also tend to finish out more slowly and may require an added 8 to 11 days to reach what is now considered a handy market weight.

Some old-school breeders have been quick to note that this slower growth curve is being played for all its worth by some show-ring pig jockeys. Such pigs may stand around a bit at a good show weight, making it possible to enter them in multiple events or "hold" them over for a particular show.

Growth rate can also impact the efficient use of buildings and lots. The quicker they make it to market weight, the quicker they move through the lots and pastures, making way for the growing or breeding animals moving up behind them. A stall-out or breakdown in a group's structure can cause a porcine traffic jam that can severely impact an operation, if not knock it out of sync for a good many months.

I almost always find very rapid days to market weight a bit dubious, while also conceding that animal reared outside will need a few more days on feed in most seasons of the year. They will be processing more feed to generate body heat and to maintain condition in harsh weather, or will have depressed appetites in very hot or muggy weather.

Back in the days of the 200 pound butcher hog, we regularly saw hogs hit the market at 4½ to 5 months of age. With 250 to 260 pound butchers, days to market have crept toward 180.

Butcher hogs, it sometimes seems, are being forced into some very unrealistic comparisons with broiler chickens. For example, a few hogs have, for short periods, posted feed conversion ratios of nearly 2 to 1, as is seen with broilers. And a few butchers once crowded 220 pounds at 120 days old.

Realistic, broad-based, and consistent performance figures will be more reflective of the hog's true nature, however. There will always be room to make them better, but we can't make them any different from the way they have been formed by nature. Better sires can be found, pastures improved, and rations fine-tuned, but there is a flesh-and-blood wall beyond which even the very best of hogs cannot be pushed.

Hogs are not the most efficient grazing animals; for them the pasturage will largely be a nutritional plus. Nevertheless, they should emerge from the experience with a muscle tone, a degree of finish, and a growth pattern that closely conforms to what the modern consumer is looking for in a meat animal.

Through dint of hard work and solid, conscientious production practices, range pork production is making a comeback, bucking trends toward streamlining and consolidation. In fact, it is a method of production that may only just now be coming into its own. There are many cost efficiencies associated with this method, not the least of which are energy savings and benefits from compliance with environmental laws.

The "dirt hog" of today is a symbol of something very different that is only just now coming over the horizon. Quite literally, it is being produced through a partnership of sorts between family farmers and concerned and informed consumers. It is a niche market with the potential to grow into something so much more.

Outdoor production is not the way to produce thousands of head per year from a single farm, but it is a way to fit hogs into a general farming mix without skewing the economic and environmental objectives of the entire operation. Our hog lots have fitted well with poultry production, beef cattle, rabbits, and other small stock, row crops, and several kinds of specialty agriculture. It has kept us in hogs when others have had to throw in the towel, and it still leaves the control of our farm in our hands. And I like that even more than I like the hogs.

PICTORIAL INTERLUDE

Purebred Hampshire sows and pigs on a good Missouri spring pasture. In the background are the classic Smidley lift-top farrowing units. One house is provided for each sow and litter.

Single sow houses drawn up to an electric source for cold-weather farrowing. Electric heat lamps can be used with no more than seven 250-watt bulbs per circuit loop. In addition, 80- gallon, two-fountain waterers are positioned to supply two different pens. When the pigs no longer need supplemental heat, pen fronts can be removed to provide pasture or range lot access.

This unit was created by the Jim Foster family to tend their purebred hogs on pasture. Its design allows feed, water and all needed supplies to be conveniently at hand during chores.

A simple, portable shade for use on range. These shelters can be topped with a ply of blackboard insulation and then sheet metal, white-painted plywood or snow pickets overlaid with straw. Modern shade material is also an option.

Good crossbred cows on range. Note the water supply line running along the top of the fence line, out of the reach of the hogs.

A shallow wallow, useful for cooling gestating and growing hogs.

A deep wallow can sometimes create problems with lot topography. Young nursing pigs have actually been lost in such wallows.

A sow doing her job, nursing her litter — and pigs doing their job, becoming hogs. Notice the clean, fresh straw bedding to a depth of 4 inches.

Freshly weaned pigs on range. Weaning is best done by removing the cows and leaving the young pigs in a familiar environment.

Young pigs exploring their own home on the range. Fresh air and sunshine do much to increase their comfort level and sense of well-being.

A few escapees — hey, it's just another part of farming! This picture also shows the service lane connecting a number of pasture lots. These lanes make moving hogs and doing chores a great deal easier.

Shade, grass and fresh air, the very best formula for high-quality pork.

A panoramic view of a well-planned range hog operation, with portable housing, easy access to all pens, and simple equipment. This setup can easily be dismantled to rotate the land back to row crops.

Water supply lines of simple plastic are run to each pen in this seasonal range operation. Growing shoats are heavy consumers of drinking water.

The sows shown here are contained in all-metal, Quonset-style buildings, which can be flipped over and slid into a pickup for ease of movement.

Feeding sows over a fence line protects the producer, as the animals crowd together at feeding time — just be sure to provide enough feed and space for even the most timid females in the group. Note the control manifold for water being drawn from the freeze-proof hydrant in the lower-right corner.

The range setup fits nicely into this landscape, with a cattle venture also situated on this farm. Swine production does not have to dominate a farm.

This young lady in classic hog-raiser garb is taking the time needed to closely monitor the development of a pen of young bred gilts. Hogs on range need to be checked every bit as closely as hogs held in confinement. Thirty minutes after sunrise and just before sunset are the best times to walk quietly and slowly through the hogs with all of your senses focused on them.

A service lane moving through a number of swine lots allows good, all-weather access for the farmer.

CHAPTER 5

FEEDS & FEEDING

Lesson one for all keepers of livestock is that you feed your mammals and fish and birds not to save money, but rather to make money.

On good pasture, hogs are able to do quite a bit to balance their own rations by consuming some plant and even animal-based proteins, vitamins, and minerals. Even hogs in drylots and wooded areas benefit from the sunshine and increased environmental stimulation. I don't think we're anywhere close to appreciating the full extent of the benefits from the odd bird egg, a few insects, a wide variety of plant life, and vitamin D laden sunshine picked up by the hog foraging outdoors.

In some parts of the country, swine pasturage exists year-round. As far north as the Canadian border, it is available for at least several months out of each year. In the out-of-doors, the hog is a complete animal, functioning best in its natural state and growing and developing to its genetic potential, free of the performance crutches that are found in confinement rearing. To me, this is a legitimate and perhaps the right challenge to swine genetics, to prove these animals worthy of being bred and widely propagated.

I am quite certain that we can't say we know everything about what a hog needs to perform and that we can provide it in a confined system. I spent 35 years in the boar trade, and I always felt better when I put boars on a truck going to an outside operation than to one where the animals would be constrained in a closed environment. I am certainly no animal rights reformer, but I just feel that the job is done best when the animals are free to function to their fullest, as the unique creatures that they are.

Pasture affords the animals a palatable feedstuff, a natural setting for life's functions, and a reduced stress load, and there's the added end result that even the fertility of the soil is improved and its tilth made better. I would also challenge the argument that farmland is now too valuable to be used for mere livestock production. Poppycock!

I like to use great old words like that one, but I have some good reasons for using it here. Some of them are:

1. Per unit of production, meat is still a far more valuable product than grain crops. Do the math and see if 10 to 20 butcher hogs sold into a premium niche market aren't far more valuable than even crop record yields of 200 bushel an acre corn or 60 bushel beans.

2. Modern confinement facilities cost tens of thousands of dollars to build, take acres out of production with the building site and lagoons, put extra waste-handling requirements on the land around them, and ultimately can be used for only a single purpose.

3. Confinement facilities and waste lagoons have little or no resale value. The lagoons themselves may even form special ecological problems that are very costly to set right.

4. Energy costs, interest costs, and overhead costs like property taxes are among the most substantial of all costs in such operations, often rivaling feed costs.

In the classic Morrison text "Feeds and Feeding," a great many pasture crop options are listed for swine production. Among the more common are alfalfa, red clover, ladino, alsike and white clovers, lespedeza, rape, grass pastures in season, Sudan grass, and cereal grasses in the spring and fall. Cowpeas, turnips, and even bluegrass have been used to provide grazing feed for hogs. Hogs on grass pastures will need more protein supplementation than if they were on legume, however.

A mix of pasture crops will work best, and the right mix really helps to extend the time on pasture. The hog can handle a great mix of varieties or even a pure stand of a single legume. The producer needs to choose varieties that will perform best in that region and with that particular soil type.

The pasture mix can be unique to the area or even to a single farm. And the choice of feedstuffs can also be used to help market the meat. It stands to reason that pork produced from hogs on alfalfa pastures and grained with an heirloom open-pollinated corn variety would have tremendous consumer appeal. An image of a primary pasture plant variety might even be incorporated into a farm logo or product packaging.

In the winter season, the animals can continue to live outside or on dormant pastures, in drylots or in wooded lots, and can maintain their nutritional needs with grain-based rations.

Outdoor production will work even on mini-farms. It will require drylots rather than pastures, however.

The drylot and the slotted platform are the outdoor options for producers with very limited land bases from which to work. Most of our young boars, for example, were grown out on slotted oak platforms that provided each animal with about 100 square feet of open-air space. With 1 inch wide slots between the 2 inch thick oak decking, the wastes fall away from the young boars. There are many plusses to this setup: they enjoyed the sunshine, the floor flexed a bit and wasn't as wearing on them, they had that stimulating outdoor environment, there was adequate space for conditioning, and they were naturally hardened and ready for life in any number of production situations.

We have successfully used drylots with our sows for many years. Our farm allows us to position sow lots on a gently sloping hillside. In each lot, we provide a minimum of 200 square feet of lot area per sow or boar. Below the lots on the slope, we maintain a 20 to 40 foot wide strip of sod. This green sward acts as a natural filter to any runoff and assures that the integrity of the slope is maintained. With this system, it is as though the lot surfaces receive a slight washing after each rain, and this has also prevented erosion problems.

Morrison held that with the pasture stocked at a rate of 15 to 20 finishing hogs per acre, you can expect savings of 800 to 1000 pounds of grain and 500 pounds of protein supplement that the animals would otherwise consume. Breeding animals, which are generally limit fed (i.e., fed a maintenance ration) to a degree, should receive grain and supplement as needed to maintain a healthful and productive state.

The maintenance ration is applied to the first two-thirds of gestation for most sows and year-round for boars in service. They should receive about 0.6 pounds of crude protein per animal per day, which can generally be achieved by using 4 pounds of a 15% crude protein ration made with yellow corn and an SBOM (soy bean oil meal) based protein supplement.

It is the task of the producer to carefully monitor the condition and fleshing of all animals on pasture and address them with proper dietary supplementation.

With breeding and lactating animals, I have always tended to view good pasturage as a plus to their regular rations. You can sometimes shave a pound of corn or 1/4 to 1/2 pound of protein supplement per day. Sows gleaning corn fields after harvest can pick up a fair amount of grain in the early going, but modern harvesting methods do not leave that much grain in the field anymore. The feed is also consumed as whole grain and is not as digestible as ground, rolled, or even cracked grains. An old rule of thumb has held that sows gleaning or hogging down a corn field would still need to be offered a couple of pounds of extra grain and 3/4 pound of protein supplement daily.

There is no Pearson's Square, no simple way to balance feeding programs for hogs on pasture or on gleaning land after harvest. It is a lot like how your mama used to prepare and serve her homemade vegetable soup: on the first day it was rich and thick and the pot seemingly had no bottom, but by late in the week she was adding beans and potatoes and was planning a trip to town for fresh fixins.

When feeding on the range, you have to walk those fields and really focus on the condition of the grazing crops, the crop residues, and the condition of the hogs. Weather is another variable to factor in, and the farther along the sows are in the gestation period, the greater the need for quality foodstuffs. As pasture or forage decline in amount and/or quality, you must bridge the gap with more grain and protein supplementation.

The hog is a true omnivore and will eat a great deal of what it finds out on the range. Anything from about 18 inches above the ground to about 6 inches below the soil surface is pretty much fair game. I have seen two sows play tug-of-war with a hapless black snake, and of course they truly relish all sorts of garden wastes. They will perform well consuming a great variety of grains that serve as the energy source in their ration. Beyond corn and grain sorghum, there is also wheat, barley, oats, and a number of food-processing wastes and byproducts that can be used as energy sources in swine rations.

The base grain in most swine rations still remains yellow corn, however. Alas, the feed-grade corn of today is far removed in feeding quality and value from the open-pollinated corn of our grandparents' day. Some of that old OP corn had a crude protein content that ran up to 13% and more. Not long ago, I saw the test results on a red OP variety that showed a crude protein content of 16%. Fed along with a bit of skimmed milk, those open-pollinated varieties made excellent finishing rations and produced a lot of ton litters prior to World War II.

In my high school vocational-agriculture (vo-ag) classes, we were told to assign a crude protein value of 9% to hybrid corn when formulating swine rations. Then it went to 8.5%, and some are now reporting tests of heat-dried corn as low as 6%. Performance has been dramatically altered in many herds, due to the declining feed value of corn.

The drought substitute for corn in much of the Midwest is a grain sorghum, milo. It can fully replace all of the corn in a ration, but it must be ground first; the best results are seen when it is mixed 50/50 with yellow corn. It will probably also need some supplementation with vitamins A, D, and E.

Wheat can likewise replace 50% of the corn in a swine ration. It is even a bit higher in crude protein content. It should not be ground too finely, however, as it can then become gummy and less palatable when chewed. Some farmers have used bakery wastes and out-of-date bakery products, also with good results. They must be tested for nutrient values to be used successfully in formulating well-balanced rations.

An acre of corn, if hogged down rather than mechanically harvested, will carry 10 to 15 shoats to butcher weight; that is, with proper mineral and protein supplementation and good quality drinking water. If they are hogging down OP corn, they will go on to produce a highly desirable, traditional pork product.

Soy Feed Supplements

The standard protein supplement is soybean oil meal (SBOM). It is made from soybeans processed for their oil and is available in a 40% to 48% crude protein range. It can be mixed with grains and vitamin and mineral pre-mixes to create balanced rations for breeding animals and most growing hogs over 50 pounds.

Another soybean-based protein is extruded soy meal. This is made from whole soybeans that have been processed with an extrusion screw (which applies pressure and heat to create an oily meal). It will not have the crude protein content of SBOM, but it does have a higher amount of feed-level fat.

These two soy-based supplements use heat and crushing to break down the enzymes in soybeans, which would otherwise inhibit full digestion and thus utilization by the hogs. Efforts have been made to breed soybean varieties that could be fed directly to hogs, thus saving much handling and processing, but nothing substantial has yet emerged. I suspect that we may never see a bean that can be taken from field to feeder with totally satisfactory results.

Another protein product is based on soybean oil meal, but with the needed vitamin and mineral supplementation added to create a complete sow feed or a growing or finishing ration when mixed with corn or other grains. It is typically marketed as hog concentrate, or "40 hog," as it normally has a 40% crude protein content.

A variety of pre-mix products are available for adding vitamins and/or minerals to swine rations. Many producers will likewise offer free-choice mineral products in either block or loose form, as an additional means of nutritional insurance. Over the years, I have busted up many a red mineral block for hogs in drylots, both to provide them with some extra mineral and to give them a root-around toy of sorts.

Starters and early-grower feeds are generally so complex in their structure that they are best bought as complete feeds. This also ensures a greater degree of freshness for some of the more specialized ingredients used in these rations. Older feed texts such as Morrison's will offer formulas for grind-and-mix pig feeds, but they rely heavily on products like tankage (dried animal residues typically derived from fat and gelatins) and meat scraps, which may no longer be available or even wise to use, in light of recent health problems such as BSE (bovine spongiform encephalopathy; a.k.a. Mad Cow disease).

Rations that are prepared on-farm are generally of the grind-and-mix or meal type. Avoid the temptation to grind such rations too much, however, since too finely milled particles can sometimes cause stomach and digestive problems. To justify the costs of on-farm grinding and mixing equipment, it is commonly believed that you need to process a minimum of 100 tons of feed per year. Quite a lot of used feed-processing equipment has been sharply discounted in recent years, and some older portable units are available at very modest prices. Just make sure that they have some life left in them and that replacement parts remain readily available.

Commercially available sow feeds and growing and finishing rations come in a variety of forms, from meal types or crumbles to pellets or cubes. The larger forms are especially useful in reducing waste when fed to animals on range. I have seen sows root 2 inch cubes out of the mud from a weeklong November rain and retrieve nearly every single one.

Processed, complete feeds offer the hogs bite-after-bite consistency that can contribute to improved animal performance. When protein supplements and grains are offered free choice, however, the animals will overfeed on the former, and gastric upset may result. Performance will certainly be reduced. Dominant animals will also consume far more than their fair share of the more palatable protein supplements.

At one time, I felt fairly confident in saying that hog feed was hog feed, whether it was bought in a blue bag in Delaware or a red one in California. The mixing formulas were fairly standard, the ingredients basically the same, and the dietary needs fairly standardized. It is a different time now, and there's more concern than ever about what you bring home from the feed store. Here's why.

1. So many of today's rations and additives are developed strictly for confinement-reared hogs, and they are simply too "hot" for outdoor hogs. The results can be comparable to trying to run a farm tractor on jet fuel.

2. The outside producer will now most likely find his or her feed needs best served by the old-formula feed products still offered in some feed lines.

3. Swine numbers have fallen off markedly in some areas, and as a result, feed suppliers sometimes tend to offer short or incomplete lines for swine. What's more, some feed products might have been sitting around the warehouse for quite some time.

4. Grain options are going to vary by region, with barley perhaps being more available in the Northwest and grain sorghum or milo more available in the lower Midwest and the Southwest.

5. Grain bought in bulk will cost less than sacked grain, but probably won't see substantial savings until it's bought in quite large quantities.

6. Processing steps like cracking or grinding, mixing, and resacking will all add to costs, and some suppliers may not even provide the service for less than 500 to 1000 pounds of feed. Some now charge substantially for delivery services or have eliminated them altogether. At the risk of giving away my age, I can remember feed charges of $5 per ton GMD—that's ground, mixed, and delivered.

7. More and more feeds and feed components are being sold in 40 pound sacks and even smaller quantities instead of the 50 pound bags that were once the industry standard. I have my own suspicions as to why this is, but the most common reason given is that with the growing number of part-time farmers, feeding chores are being done more

and more by women and youngsters. I grew up during the last days of feedstuffs in cloth and burlap bags and can still recall 100 pound sacks of bran and a few other feedstuffs. It was quite a benchmark in a young man's life when he could lift one of those monster bags free from the loading dock and carry it onto a truck.

8. There is often a surprising inconsistency in feedstuff prices and availability, even among local suppliers. Our two area elevators may sometimes be as much as a dime a bushel apart on corn prices, and one will stock milo all the time and the other never. I've seen up to a $2 to $3 per bag difference on swine feeds at some area dealers. Just be very suspicious of exceptionally low price; there is often a very fine difference between those bargain-basement feeds and warehouse floor sweepings.

As noted earlier, I worked for some time in the livestock feed business and saw many things there that would chill you to your toes. It was there that I learned that one definition of an expert is anyone more than 50 miles from home and able to speak in a loud and forceful voice. To illustrate what one might encounter there, let me as a survivor of the great "hog finisher war of '77" relate my war story.

Corn was cheap that year, and our company came out with a 13% hog finisher that sold for a bit over $80 a ton, sacked and delivered. We sold a good bit of it until the Illinois soft wheat crop was hit hard with wild garlic contamination. One of our competitors came up with a wheat-based, slightly garlic flavored finishing feed that they sold for under 70 bucks a ton. We went to $65 on the boss' orders, and they trumped us at $63 a ton. Short of telling each buyer that I was the illegitimate son of their field rep and had been abandoned as an infant, we had nothing left on which to compete. They sold a lot of their garlic-flavored finisher that season, and for a good many years afterward area farmers watched their wheat crops closely, anxiously for any resulting garlic intruders.

Feedstuffs are generally tested by state departments of agriculture for quality and truth in labeling. They aren't supposed to be resacked or sold retail in damaged bags. Here in Missouri, I have seen ag department agents visit area feed stores and pull samples or entire bags of feed for testing. In addition, the ration's contents are supposed to be listed in descending order of amount on the sack's tags, along with the feed analysis, feeding instructions, and any notes concerning additives and withdrawal times.

Choosing the Right Feed

Hog feeds vary greatly in content and form, but beyond the inconsistent protein content found in modern hybrid corn, most branded swine feeds are fairly reliable.

Some are computer generated to create a least-cost formula and thus may vary in content rather markedly from week to week. This can affect consumption and performance, but generally such changes are designed for minimal impact. It does explain, however, how the color and aroma of some feedstuffs can vary from ton to ton, if not from bag to bag.

Most companies do offer feed lines that they guarantee to remain consistent in their formulation, but they also tend to sell for higher prices than the computer-generated rations. Still, in only a very few instances have I ever encountered a truly inferior feed product. Some can be misused or fed to animals in the wrong age or class, but there are few—very few—truly poor feed products to be found now.

There is a smattering of somewhat generic lines that have been put together on the cheap for sale in some general farm supply stores. These may fail to generate desired performance levels in some instances. You definitely want to beware of the so-called all-purpose or all-stock feeds. It is hard to go wrong with name-brand feed products, although sometimes you may have to pay a bit extra for the label.

Look for feed products from firms still doing research and developing new or upgraded feed products. You don't always need the latest new thing, but such practices do attest to a company's commitment to quality and customer success.

A good feed dealership will have someone on staff or have access to company personnel who can field producer's feed questions and can help formulate a feeding program. The range producer will be very poorly served by rations formulated for animals held in rigidly controlled environments, and may have to connect bits and pieces of different feed lines to create the optimal feedstuff.

The range animal will need additional energy in cold and inclement weather, and perhaps a more nutrient-dense ration in warm weather, when the heat and humidity may reduce appetite and consumption. Range rations can be a bit simpler in their overall formulation, however. The range producer will fuel his or her animals with rations built to undergird them while on pasture and bring about optimum, profitable performance. To that end, read those feed tags to make sure that your animals are getting what they need and that you are not paying for something your animals really don't need or can't utilize.

Comparison shopping for feedstuffs is a must. I value one of our local elevators very much for the simple fact that they allow me to freely roam their warehouse, reading feed tags to my heart's content. They work hard to keep their supplies fresh and are willing to access all company resources to help their customers.

As long as quality product is a central element in your production system—this should of course be a given—always balance price considerations with such services and assistance your local suppliers make available to you. There are feed stores where the standard answer to every question is to hand you one of those little feed company brochures. If you don't find your answer there, well that's just too bad. I've also been in farm supply stores where I knew more about their product lines than the people on the floor and behind the counter. They hadn't even taken the time to read their own press releases and advertising. Making well-informed decisions, even if it means paying a little extra to someone you can trust or who understands your specific needs, is crucial to maximizing your farm production and minimizing the mistakes.

Feed Storage

If you can provide the needed on-farm storage at a reasonable cost, there are some advantages to buying in volume and making seasonal purchases of feed grains. Here in the Midwest, there are generally favorable grain prices at harvest and during off-peak periods for livestock feeding. However, with a short crop, the highest prices may be seen at or shortly after harvest.

Volume discounts for feedstuffs generally kick in modestly at the 3 ton range. There are also occasional preseason discounts for protein concentrate purchases. Buying in bulk will certainly save sacking and some other handling costs, but you have to be prepared to receive, store, and distribute bulk feedstuffs in ton-plus lots.

A pickup truck loaded with 5 gallon buckets holding various kinds and amounts of feedstuffs is still a fairly common sight in our part of the world, but there are other good, inexpensive ways to handle feeds and feeding on range. If I had a dime for every yard I've ever hauled a 5 gallon bucket of feed, I'd have one fancy pickup with a whole bed full of dimes. Over the years, I've seen a lot of feed stored in the beds of old straight trucks and wagons pulled into sheds and under lean-to's, out of the weather.

In my younger days, we were one of the last in our area to continue to pick ear corn, and honestly, I have yet to see any way to feed whole corn to breeding hogs on range better than pitching ear corn over the pasture gate. Seventy good ears make a bushel, and I swear Dad could pitch out ears to assure each and every sow in a ten-head drove got the exact number she needed to keep her slick and shiny. That first morning after harvest, with the bins filled to the brim with ear corn, and the air rich with the golden summer smell of grain, was always a very special moments for us.

Over the years, we reworked that old granary into many different configurations. With reinforcement lumber and plywood sheathing, we converted some of those old ear-corn bins to shelled-corn and other grain storage, but Dad always spared at least one slat-fronted bin for his favorite kind of corn.

Modest but handy feed bins can be made from old round swine feeders. On most of these sorts of feeders, the bottoms fail long before the throats and bins do, and they are fairly simple to remove. To convert one of these old feeders to a 40 to 80 bushel bin that can be easily positioned around pastures and lots, follow these steps.

1. Carefully inspect the tank and throat or cone to ascertain that they are free of any rust or other damage.

2. Cut a circular piece of 1 inch treated plywood to fit into the small end of the feeder cone. Affix it in place solidly with screws.

3. Into the plywood disk, cut a modest 4 x 4 inch square hole. Cover this opening with a simple sliding door. Sometimes sliding grain doors can be salvaged from old elevators or grain trucks. I once bought six sliding metal grain doors for a bit over a dollar apiece at the local sale barn.

4. Four uprights should be bolted or welded to the sides of the tank to serve as legs for the bin. You can use angle or channel iron or treated 2 or 4 inch wooden stock. I like the supporting legs to reach from the top lip of the tank to well past the end of the cone to allow a 5 gallon bucket to be positioned beneath it when set upright. The buckets should slip in and out beneath the cone easily for filling with feed when the sliding door is drawn open.

5. To the bottom ends of the support members, attach 4 x 4 or 4 x 6 treated runners so that it can be easily towed about the hog operation.

After bottom failure, the most common reason for discarding a feeder is loss or damage to the top lid. Wind can really play hob with these, and many farmers have a bad habit of filling the feeders too full and leaving the tops thrown open until the feed sifts down. Most feeder companies offer replacement lids for their feeder models at fairly modest prices. You can also scavenge most large hog farms and come away with large numbers of lids and other feeder parts from old feeders that have been discarded there. The fencerows of Midwestern farms are legendary for the recyclable treasures they can yield up to diligent searchers.

Another even simpler feed storage unit can be made from a discarded chest-type freezer unit. These make waterproof units that will hold several hundred pounds of feed and that are fairly easy to move about once the motor and compressor have been removed. They are generally free for the hauling at area appliance dealerships, or else you can simply put out the word that you are looking for some, and before long the phone will begin to ring.

A critical safety measure, of course, is to disable the latching mechanism as soon as you get it home. Most of these freezers will hold several 50 pound bags of feed set on end, and some can be partitioned with simple plywood dividers to create two or three mini-bins.

These old freezer cabinets can be set on simple sleds or runnered platforms made from treated or native lumber and can be pulled adjacent to pens or pastures as needed. An occasional coat of a good enamel will keep them looking bright and clean. We have one that has been in steady use for nearly 20 years, and it is still keeping feedstuffs clean, dry, and fresh.

The Producer's Daily Task

The most important feeding tool the producer ultimately has is an experienced, diligent eye. He or she must remain steadily focused on the health and conditioning of all of the animals. My father always said that the good ones knew what their hogs needed even before the hogs did.

For example, gilts going into the breeding program need to be carefully monitored. As a producer, you will be helping her to develop her frame, reproductive system, and her first litter of pigs all at the same time. She will need at least 1 to 2 pounds more of feed per day than her older counterparts in the breeding herd. In the last third of the gestation period, a time when fetal growth is greatest, you need to be watching her progress on a daily basis. If she falls behind here, she may remain behind the rest of her life or fail to remain in sync with the herd and have to be culled after her first litter.

We had a Hampshire gilt farrow deep into an August hot spell, and her feed consumption fell off to nearly nothing. Even having weaned her pigs at just 3 weeks of age, she suffered a loss of weight and condition that she was never able to fully build back up. I would say it put her behind a good 100 pounds for the rest of her life.

Gilts developing on good legume pasture have a very palatable nutritional extra to draw upon. It is a resource that cannot be easily measured, but when in place, it will do much to keep all animals in the herd properly supported nutritionally. How many mouthfuls of good greens a gilt or sow needs to stay in position in the herd just can't be calculated—it's based on keen observation—but without them, the producer must fill the void with rations not only well prepared but metered out to the exact needs of each individual animal.

The outdoor sow is in her element. In the wild, she is generally the big mama of the woods, and in the open air she is free to function to the fullest extent of her naturally ingrained traits and temperament. Hogs are marvelous creatures, curious about the world around them, naturally equipped for both varying terrains and seasons, and among the most pleasant of creature with which to work if not too oppressed and constrained.

Many years ago, Dad was doing morning chores by himself when he got caught in the rush of a set of hungry sows. A glancing blow to one of his knees caused it to lock up on him, and down he went. The senior herd boar was with that group, and when he came close, Dad draped himself across his back. The boar then slowly made his way to the lot gate, where Dad was able to brace himself and regain his bearings. While I don't believe that old boar was Lassie in a pig suit, he was clearly an animal that had been handled gently and treated with some respect in his time on our farm, and that kindness paid off.

Hogs are social animals that function best in group settings. Sometimes it seems like they all want to eat or drink at the same time. In environments where stress from unnatural conditions is reduced, they are free to relax and do all of those little things that make them optimally productive. Everything from simple creature comforts to trace nutritional needs to vital sensory stimuli are accessible to them in an outdoor environment.

Aside from land costs, the next big argument against rearing hogs out-of-doors is that there they eat more. It's true they do, but feed is a totally deductible expense in the year it's eaten, and the amounts consumed compared to indoor hogs aren't all that much greater. Days to marketable weight may be extended by a few, but inclement weather can also impact hogs held in many confinement units. In cold weather, hogs need extra BTUs, either from feed or supplemental heat. Both have a cost, and experience shows that the farmer now can do more to manage and pare feed costs than energy costs.

The classic farm texts say that a sow in good condition will meet all of her daily nutritional needs with 4 pounds per day of a 15% crude protein ration, or 3 pounds of shelled corn and 1 pound of a 40%

complete supplement. And in an ideal world, Raquel Welch would have been my prom date.

On pasture, we are moving closer to the hog's ideal world, however. Things are a little more forgiving out there than on a sheet of concrete or sitting atop a pit of stagnating wastes. If nothing else, the air exchange rate out there is in God's hands, and not the electric company's.

One object of neglect is sometimes the herd boar; even though he is the lead animal in a group, his nutritional needs may sometimes be slighted. For example, a few years ago we would occasionally see some breeding boars with chipped toe points and some fine checking in the hooves. Some breeders maintained that it was due to a biotin deficiency, and others scoffed loudly at that assumption. Soon after, however, we were seeing biotin being promoted as a new, important addition to breeding stock rations.

In most instances, the boar will be well served by the same ration that is fed the females. When he is not needed in service, it is best to separate the boar from the sows and keep him on a simple maintenance ration. We try to keep a small, well-shaded lot for this purpose, some distance from the females.

If not monitored carefully, boars can soon grow too large to be used safely with gilts and some smaller sows. In an earlier day, when hogs were bred for lard type, it was quite common to see breeding crates on farms with hogs. The female in estrus would be placed in the crate, which would be mounted by the boar and bear his great weight during the act of breeding.

We once had a Duroc boar that seemed to grow on little more than air and water. We had ol' Jack down to naught but a few ears of field corn each day during good weather, and still he thrived. Another time, a young Red boar was loaned to an elderly neighbor who, to use a good Missouri expression, "fed 'em with a scoop shovel." Over a couple of winter months, he actually did little more than throw feed over the fence to a handful of hogs and then hurry back inside. In short order, he called us to pick up a 16-month-old boar that crossed the scales at the local buying station weighing 745 pounds.

Farrowing & Lactation

At farrowing, nutritional care and overall needs change rather dramatically. For the first 24 hours following parturition, the sow will eat little or nothing and then should be gradually brought up to full feed. Full feed is the amount of feed consumed daily, roughly equal to 3% of the animal's live weight. On a 400 pound sow, this will be 12 pounds of feed per diem within 7 to 10 days following farrowing.

An old rule of thumb holds that the nursing sow should receive 3 to 4 pounds of feed each day for herself, and another 1 pound for each pig she is nursing. Here again, the important thing is to build gradually to this level, to prevent any gastric upset that could adversely effect milk flow.

Essentially, the lactation ration is the same as the gestation ration, just multiplied two or three times over by a week or so following farrowing. Some will, for a day or two following farrowing, feed a ration liberally laced with a source of added bulk, such as bran, oats, or beet pulp. This is to get the bowels open anew and to prevent any milking problems triggered by constipation.

Over the course of a year, a sow will eat roughly one ton of feed. If she's on good pasture, as much as 500 pounds of this feed can be trimmed. The sow that is free to move about will have better appetite and consume more, and this should translate to better milk production and pig rearing.

Water needs will amp up significantly for the nursing sow. Water is necessary for digestion, milk flow, and a great many other body functions. There is no available guide to consumption levels, especially since this will vary with environment—air temperature will certainly be a factor—but figure a minimum of 10 gallons of water per day for a nursing sow. If she is watered in a trough or pan, you may need to offer the sow fresh water as many as three times a day.

Likewise, to reduce waste and get as much feed as possible into a nursing sow, you may have to offer her fresh feed two or more times a day. I've had some Chester sows that needed 18 to 20 pounds daily in the latter stages of lactation, but they simply couldn't consume those amounts in a single feeding. I also like to see those sows get a good feeding and drink of water going into the nighttime hours, particularly on those long, cold winter nights.

Just because your sows are on good pasture doesn't mean you can neglect them and their conditioning. You can't just dump them out there and then go along your merry way, believing all will be well and good because you filled the feeders and topped off the water tanks. An awful lot of good hogs in the Midwest still die from "corn planting"

Postnatal Care of the Sow

The weeks just before and after weaning is a stressful time for a sow and her pigs, so major changes that might affect the animals should be avoided if at all possible. During this period, she should be eating substantial amounts of feed, living in comfortable surroundings, and should emerge from the nursing period thinner but able to quickly rebound and regain her prior condition.

To be sure, she will have nursed down and will be lighter in weight than when she farrowed, but not so much that she will be in a severely depleted state. You want your females to be able to be back breeding within a couple of weeks following weaning in order to keep your breeding herd on schedule. Many farmers will flat-out tell you that the female that is neither bred nor nursing is costing you money with every breath.

Many years ago, we had a set of nursing sows very nearly starve to death while being amply fed what was supposed to be a complete grind-and-mix ration for sows. They were older sows with several good litters behind them, but were large framed and had developed well. We noticed that handfuls of the ration felt light and wasn't the right consistency for what was supposed to be a rather nutrient-dense feed product. Its dusty composition led us to suspect it had been made with old ingredients, and there appeared to be substantial amounts of some sort of bulk product present in the blended ration.

A feed sample retrieved from a back corner of a bin pretty well confirmed all of our suspicions when submitted for lab testing. Unfortunately, the results arrived too late for us to make any sort of formal claim or complaint. It has taught us that pulling an occasional feed sample and storing it away in a glass jar with the date and the particulars is a good idea. And especially so if you are changing rations or suppliers.

On many farms with hogs, you might see jars on posts—a jar lid will be affixed to a gate post, and a jar is then screwed onto the lid. This serves as a water-tight container for storing field notes and the materials for taking those notes. When the animals are moved from pen to pen, the jars can be quickly unscrewed and their field records moved right along with them.

If the newly weaned sows are exceptionally thin, keep them on their increased feed rate for a few more days. With sows that have been severely affected during lactation, it may be necessary to drop them back an entire 21 day breeding cycle.

When you see a female rapidly losing condition, there are a number of things that can be done. You can early wean a litter if they are vigorous, performing well, and can handle solid feed. You can also boost the sow's ration with added grain and protein supplements, but do this gradually. An old purebred producer showed me a trick of boosting a nursing sow's ration by adding 1 or 2 pounds of nutrient-rich pig starter daily. It isn't cheap, and the feed has to be gradually built up over the period of a few days, but it will bolster a sow's nutrient uptake quite well.

disease. Pasture production does not free you up from the tasks and responsibilities of a good stock raiser.

Growing/Finishing

Growing or finishing a hog on pasture, in lots, or in harvested grain fields is not simply a matter of "turn 'em in and let 'em go."

That being said, let me begin with a locally renowned anecdote. The county just to the north of ours has led the state of Missouri in pork production many times over the years and has spawned a number of colorful hog men and hog tales. A good friend of mine who worked for one of those gentlemen related the following true story.

Late one fall, they brought in a set of sows and pigs from a remote wooded lot on the producer's farm. My friend heard a crashing through the brush as they were bunching and loading the sows and their good-sized pigs. The owner dismissed the noise as "varmints," and a few minutes later they shut the gate on that lot for the whole of the winter season.

The next spring, my friend was sent back to ready that lot for another season of use. In the lot he found a small set of shoats that had survived the winter on the feed left in the feeders, mostly, and whatever else nature brought their way. They were far from market ready, but they were alive and well, and that says a lot about hogs and the natural life. Just imagine what they can do with good care.

Growing and finishing hogs are not the most efficient uses of pasturage, but they do glean grain fields well. Drylots offer them the stimulus and other life qualities of an outdoor environment that are so conducive to good swine performance. You will not achieve test-station levels of performance out-of-doors in simple facilities, but you can come close and at quite favorable costs to produce.

As mentioned earlier, the outside hog will consume more feed in some seasons to counter the cold and dampness, but not that much more. And trying to offset feed costs with building and energy costs just doesn't make a lot of sense to me.

Furthermore, the outdoor-finished market hog would appeal to the same premium market that has grown up around range broilers and natural beef. This is the humane way to produce pork, and the real plus is that it is also the best way to produce pork.

Taste is a function of many things, ranging from texture to aroma to eye appeal to personal expectations of actual flavor. Follow a semitrailer load of confinement fed hogs very far down the road and you will encounter an odor vile enough to turn even the staunchest Texan into a vegetarian. To informed and concerned consumers, range pork is the real deal. To older consumers, it is the pork so well remembered from their youth, and it is the family farm product that deep down we all really wish for.

The traditional practice in hog rearing has been to feed the butcher hog a ration with a declining protein level as the animal grows nearer to a good harvest weight. You would begin with a 15% to 16% crude protein ration and continue with it to a weight of about 125 pounds. From there to about 175 to 200 pounds, you would feed a 14% ration. Then you would finish them to the desired market weight with a 12% or 13% crude protein ration. It worked fairly well and generated some cost savings, but swine genetics have gotten better over time, and now you are growing muscle rather than adding to fat cover.

Even with the increase of handy market weights into the 240 to 260 pound weight range, many now feed their hogs the same 15% ration throughout the entire growing/finishing period. Such a ration will accelerate the growth rate a fair bit and generate an even leaner carcass.

About 10% of a growing hog's nutritional needs can be met by good legume pasturage. It is less utilized by the animal in the finishing stages of the growth arc, but still it is not a waste of resources. Hogs finished on range emerge as fitter, more toned meat animals. They may perhaps yield pork with a bit more fat content, but its table qualities will be unequaled.

Some studies now are actually beginning to contradict the above assumption. They are showing that hogs finished "in the dirt" may actually be hanging carcasses of superior quality compared to some confinement-finished hogs.

With butcher hogs on pasture, you will be confronted with three main challenges from the swine industry. They are that (1) you can't do it with land priced as it is currently, (2) they won't perform up to their potential, and (3) the costs of the final product cannot be justified.

Piffle! There is another one of those great old worlds that we don't get to use often enough but do sometimes fit the moment like no other.

Hogs fall into the livestock category of "large animals." Lord knows they've tried to raise them in stacking cages like rabbits, but the truth is that you are only going to get them to live and grow if you spread them out. And that means either penning them in sheet-metal gulags or in rotating lots and pastures that can be turned back to crop land. How a near million dollars worth of single-purpose buildings or acres of concrete pare land costs has never been made clear to me.

Nor can you raise thousands of hogs outside on a single farm. There is, however, a most telling specter that hangs over pork production here in Missouri, and all across the nation actually. Had the family farmers in place along the Iowa-Missouri border taken up sow herds of 50 head or less, they could have met the pork market emerging there and staved off the corporate giant, Premium Standard Farms, that is now so firmly entrenched there.

Ten acres, even at $4,000 an acre, is still an investment far less than the costs of even a single modern finishing house. It will not deteriorate and depreciate like a building, it will always have multiple uses, and it will continue to keep you in the role of farmer and not hog house technician.

As for performance, there is now strong evidence that range-farrowed pigs have a lifetime advantage of hardiness and vigor. Hogs bred for the dirt don't have to take a back seat to any of their porcine fellows. And as noted earlier, some performance data now is as much artifice as it is honest.

Feed 'em creep feed at a constant 70 degrees, and you'll step up their growth, but at a cost that current selling prices simply cannot warrant. And in most head-to-head studies, the differences they are citing are measured in differences of mere pounds and days. On most range operations, those constant overhead costs are not nearly as high as they are for the confinement producer. A statistic that I would like to see but have never seen published anywhere is survival rates for hogs fed outside and in confinement.

As for the end product? Well, the folks that have been vocal in their criticism are those whose job it is to see just how few chocolate chips you can put in a cookie and still call it a chocolate chip cookie. Food retailers in outlets as diverse as fast food stops and upscale markets are making it quite clear that they and their customers want pork of substance—nutritional, environmental, and moral substance.

People who remember their grandparents' pork as being somehow better aren't nostalgic or delusional. It was, to paraphrase the tag line of that old Carnation milk ad campaign, "the pork from contented hogs."

Feeding

With finishing hogs on pasture, position your self-feeders as far as possible from the sleeping sheds, to encourage more muscle-toning exercise.

Keep a good 14% to 15% crude protein ration for the hogs on pasture at all times, to assure good growth and to prevent any performance-robbing gastric upsets. You will need to provide a feeder lid for every five to eight head of growing hogs. If the pasture quality continues well and the animals are growing well, the protein level might be pared a point or two late in the finishing period.

A pelleted grower/finisher will trim some wastes, and such processing will free up more nutrients in the feedstuffs, but this is a practice that may not always be cost effective. Some feed suppliers also offer pelleted feeds at bulk purchase rates, beginning at 3 tons or so. This is the amount that will generally fill one bin on a bulk delivery truck. The more you can shape your feed needs to the industry's ability to supply them easily, the more you will be rewarded with price savings.

Normal feed performance with today's better swine genetics is 4 to 4½ pounds of feed per pound of gain. Old hands will delight in telling you it takes 10 bushels of corn and 100 to 125 pounds of protein supplement to produce a "fat hog." Over the years, I have seen some swine-feed efficiencies crowding the 2:1 ratio long held out for broilers, but they were achieved under far from real-world conditions.

There is no hard and fast rule that says if you feed X amount of feed you will get back Y amount of pork. Feed efficiency and growth rate are among the traits with the greatest levels of response from genetic selection. Still, you can't breed your way through a truly bad environment or climate extremes. Extreme heat and cold will slow them down, and too much muck and dampness will bring them to a standstill.

In the classic drylot situation, the hogs are out-of-doors full-time, but the spaces are reduced, and there is no browsing or pasturage. We have used drylots for decades for both our breeding herd and growing hogs. For the producer with a very limited land base, they are the way to go; the hogs produced there will enjoy many of the same benefits received by a pasture-reared animal.

In such lots, you will have to provide 100% of the nutritional needs of the animal. In our system, we work first with a simple cold nursery that consists of an open-fronted house pulled adjacent to a platform of the same dimensions. We wean the pigs into this type of unit and in this house gradually shift them from a starter to a grower ration.

I like to keep them on a 15% ration from the time they weigh about 50 pounds up to market weight. When growing out boars for sale, we keep them on a full 15% ration until they reach 300 to 350 pounds. Young boars beyond this point would then be fed this ration in limited amounts daily until they are sold.

Self-feeders in drylot also need to be set on platforms that extend past the feeder base at least 2 feet on all sides. You may have to move them often about the perimeter of the lot to prevent mud buildup. In some drylot situations, an investment in concrete or gravel pads to server as feeder/waterer bases may be justified.

I favor the smaller, 40 bushel rounded feeders for their stability and ease of handling. Also available for use along the fenceline are single and double sided rectangular feeders of similar capacity. Bigger feeders are harder to handle and position, and many seem to be top-heavy. Feeder lids should always be closed down and well secured. It is an all-too-common practice to top off self-feeders and leave them with their lids pushed back until the feed is eaten down. If you're in desperate need of a big rain, do this; otherwise, be sure to seal it up.

I have had to clean out a few plugged up feeders over the years, and it is not one of my favorite things to do. Ideally, the last bite of feed in a feeder will be consumed just before the last hog in the pen goes up the chute. That seldom if ever happens, but it's what you shoot for. You don't want to have feed sitting around in a feeder for very long.

In Missouri and many other places, we still see a fair number of hogs in wooded lots. This is really back home again for hogs; they really seem to take to our flinty Missouri hillsides. I would not run hogs into true woodland, since they can take a toll on valuable trees and some delicate forest soils. In scrub, there is still a lot of vegetation to be gleaned, and they can clear land on a par with most brush goats.

With good fencing, hogs can be positioned on any odd parcels around the farm. We have used hogs on a fescue stand to take it down a bit and make it possible to introduce new plant species into it.

It has been rightly said that every hog is born with a bulldozer on one end and a manure spreader on the other. With a bit of planning, these traits can be put to use, or at least mitigated. The reputation hogs have for creating a bare environment in their wake is actually undeserved.

Pack them in too tight, without a plan for rotation and proper management, and you can have problems. There can be runoff from some lots, hogs can kill some young trees, wastes can build up, and there may be some local odor problems. This is not a condition inherent to the hogs themselves; rather, it is the result of too many (likely neglected) hogs being kept in one small space.

When a wildlife population becomes too great for a given area, natural checks crop up (often appearing as health and reproductive problems) to restore things to a more natural balance and order. Winter deaths will be greater, health problems like mange and rabies will emerge, and habitat will begin to break down. If only those producers who are working in confinement units and constantly battling the diseases of the month would just step outside long enough to learn this simple lesson that nature has to teach.

Even the smallest of farmers have had bad problems trying to push too many animals through their facilities. When this happens, many hogs simply die from what is best termed "too many hogs disease."

Drinking Water for Finishing Hogs

Water is the most important of all feedstuffs, and providing it in sufficient amounts on range can often be a problem. I have packed my share of 5 gallon buckets of water over the years, and nothing is more exasperating than trundling water through the August heat and then watching as a hog takes a couple sips and then upends the whole trough.

I have seen hogs allowed to run freely to ponds and creeks or streams, but good pond management and environmental concerns now preclude these practices in most instances. A growing hog will consume a minimum of 2 to 2½ pounds of water for every pound of feed consumed. That water should always be as clean and fresh as possible.

Water can now be piped fairly easily to temporary pastures and lots, and even to some grain fields as they are being gleaned. To the extent possible in the planning and layout stages, try to use gravity to simply and inexpensively move water to where it's needed.

Simple, typically aboveground plastic piping, modest pumps, and gravity can move around a lot of water in a seasonal program. A 250 to 500 gallon tank and a trailer made from an old pickup bed will likewise cut down on the need for packing water buckets.

I've fought to get a hog or two up out of binding mud, and thus the need for its control and prevention have been made abundantly clear to me. To prevent mud buildup, move those waterers often, don't slop water when filling them, and inspect them frequently when in use.

We've done a lot of watering in troughs, but with these, hogs have the nasty habit of taking a sip or two and then thinking "bathtub!" The first line of defense was to weld a pipe or rod down the center of the trough. This wasn't such a good idea; it's a good leg breaker when the trough flips up. Alternatively, cross-members welded to the trough just gave them something to lie atop while they dangled their tootsies.

The trick we've learned is to slide just enough of the trough under the fence to allow them to drink from it like a fountain. Stake it in place with a short rod or pipe driven into the ground on each side of the trough. This works even under hot wire if you give them just enough trough space to drink from and if the trough is pegged down solidly.

When we first began raising hogs, a line of heavy iron drinking fountains could still be found. They could be mounted through a barrel wall at ground level to create one- or two-sided hog waterers that were easy to move about and inexpensive to own. They would outlast a score of barrels and a ton of abuse. Alas, they have joined the hog oilers as curios for display in the parlor rather than continuing as a useful, everyday tools.

No two farms on this Earth are ever exactly alike. The question of how best to handle water will have a different and unique answer for every farm. For cases in which money is no object, you can bury miles of underground pipe, pour concrete pedestals, and place heated, automatic waterers in every lot. On the other hand, you can get by well enough with troughs made from halves of old water heater tanks and fill them with 5 gallon buckets as you make an extra effort for your physical (and fiscal) fitness. Somewhere in between is the solution that will work for most farms, but even with the ideal setup, you will never be totally free from knee-boots and plastic buckets.

The Way We Were — The Way We Can Be Again

I recently heard from an extension agent who had relocated from the Missouri mini hog belt to the tall-corn state of Iowa. Much to his surprise, he discovered there a great number of sow operations with 50 to 100 head that were mostly outside and doing quite well. We were both surprised; those aren't supposed to be there anymore.

The hog raised outdoors has found a niche—a quite fair-sized niche—that is uniquely its own. It is, however, no longer the conventional hog.

The "dirt hog" is now fed and grown for the market through the learned experiences of conscientious producers and informed discernment on the part of the consumer. (As an aside, the term "dirt hog" is a rather odd but not uncommon pseudonym for these animals, one that would not seem to create a very flattering image, but more and more, it is being used to describe hogs of merit and value.) On range, you are feeding for optimal rather than maximum performance. Yes, pounds of gain are a concern, but no more so quality of production and the degree of consumer acceptance of that product.

We feed hogs to make money with them, and the feedstuffs are their fuel for growth and reproduction. On pasture, they can do a fair amount toward balancing their nutritional needs. In a drylot, they are bathed in the sunshine's goodness and have a higher degree of naturalness and contentment than hogs held in rigid confinement.

The old dirt hog is being made new again and has returned as a real contender on the farm, on the rail, and in the retail marketplace. We are still fairly low on the learning and development curves for this type of production, however. This is because the practice was abandoned too soon in the move to confinement rearing and has gone for at least a third of a century with little or no input from researchers or the feed industry.

DUROC JERSEY

MARKETING

We concluded the last section with a brief note on value and merit, the traditional one-two punch for successful marketing.

At the local buying station or terminal market, all hogs may not look the same, but they are essentially valued the same. These are the market terminals for a wholesale market, with virtually no reward for anything but volume. For nearly as long as I can remember, the packing industry has professed a desire for animals of exceptional carcass merit and yield, and yet little if anything in the way of a price premium has been offered up for it.

In fact, when prices turned down steeply in the mid-'90s, some elements in the packing industry effectively punished farmers financially if they failed to produce a certain animal type that was favored not for any obvious merit to the consumer, but because it processed out in a way that pared packing-line costs. To put it as succinctly as possible, pork as a mere agricultural commodity is a no-win situation for the independent family farmer.

Virginia ham and lean, free-range pork are not pork as a mere commodity. They are even more distinct than simple branded products. Well-cured ham is Christmas mornings, the warmth of your grandmother's kitchen, Robert Frost poems, and, well, history and home on a good china plate.

The "pork industry" has determined that pork is to be bought as cheaply as possible and marketed with all of the imagination and blandness of a 29¢ box of animal crackers. The "pork nugget" has yet to arrive, but as many believe, the pork industry will not be happy until hogs are delivered to them boneless, skinless, and with little dashed marks outlining the prime cuts. That day may not be far off.

The outdoor producer who has assembled a herd constituting something more than mere "commodity pork" is going to have to strike out on his or her own to find the markets that appreciate and are willing to properly reward those efforts. It begins by looking well beyond the traditional hog farm end product—250 pounds of rompin', stompin' porcine. That hog is the kind that's all too quickly handed off to the packing industry at the nearest buying station or sale barn in a process that keeps the producer and the consumer a very great distance from each other.

The Butcher Hog

The hog as a meat animal was regularly sold off the farm at a market-ready (handy) weight from lots that would keep one self-feeder fully utilized. Ten to 15 would fit in the bed of a pickup and were fairly common-sized lots at the interior buying stations.

In my lifetime, butcher weights have progressed from a 220 pound ideal to the current 245 to 265 pound desired weight range. The lightweight hogs are now rather sharply penalized in price, while butchers up to 300 pounds have found acceptance. The packing industry seems resolved to make the butcher hog little more than a block of meat that improves output from the processing line while paring labor costs.

They have, I believe, chosen to ignore the merits of the lighterweight slaughter animal. With the 220 pounder, there are fewer pounds of product to depress the market (ten 250 pounders have greater impact on that scene than eleven 220 pounders). Furthermore, the pork is from younger hogs and thus they are more tender—as hogs mature, both growth rates and feed efficiencies decline—and the larger hogs take a greater toll on finishing facilities. These bigger hogs are on the farm for a longer period, consuming more feed and ultimately producing larger primary and secondary cuts of pork, which would appear to be at odds with today's trend toward smaller families. In parts of Great Britain and Europe, there is a tradition of harvesting meat hogs at weights as low as 110 pounds, to produce a very lean and efficient variety of pork for roasting.

As we ponder these super-sized butcher hogs, we should perhaps bear in mind one humorist's definition of eternity: two people and a ham. It is one thing to still be eating sandwiches made from the Christmas ham on New Year's Eve, but now you might not finish it up until Super Bowl Sunday.

The pork industry has failed to develop a fast food or convenience product comparable to the chicken nugget or chicken tender. Such products make possible the maximum utilization of poundage from such monstrously proportioned creatures as the modern broiler. The two restructured products that the pork industry can point to are the riblet and the pork burger.

The former product has been taken up by at least one fast food chain and is used occasionally in the food service industry, but it will never compete against the hamburger or even the hotdog as a sandwich favorite. The pork burger has found little acceptance outside of the Midwest, although it can be made with either carcass trimmings or in the whole-hog style. It will do everything hamburger will do, except of course sell by the millions through the fast food trade.

The premium market for pork still lies with the classic pork products: ham in all its forms, bacon, thick pork chops, and that other breakfast standard, pork sausage. Everything from tails to feet to ears and snouts is eaten somewhere in this country and has value. In a restaurant tour of St. Louis, you can buy inch-and-a-half thick chops that will rival the finest steaks, mom and pop ethnic places that serve "snoots," and others that feature that blue-collar favorite of the Midwest, the barbecued pork steak. Some cultures favor roasting pigs, and others can do a lot with even the biggest of sows. Older boars moving out of breeding herds often are shipped east, where they are processed into a number of spiced and distinctive flavored sausage products, such as pepperoni.

Marketing Then & Now

Over time, I have learned that the most successful swine producers are also first-rate art collectors: they cherish and amass green portraits of dead presidents. They have learned that success is measured not by how much money you get to handle during the course of production, but by how much you get to keep after all—I repeat all—of the bills are paid.

They park the tractors and step into the role of businessmen and women, because that is what is needed now to survive and succeed. They have seen a very troubling light at the end of the tunnel, one that dictates a sea change for production agriculture.

For decades and even generations, American livestock production has functioned with a fairly vast and complex marketing infrastructure. There were real "butter and egg" men who, from an East Coast base, moved throughout farming country, pulling the farming riches of the plains and prairies into the cities, big and small.

We grew up relying on order buyers, commission men, terminal markets, sale barns, and buying stations. On Sunday evening, we could ship hogs to the National Stockyards at East St. Louis, on Tuesday evening we could run an old boar to a nearby buying station, and on Wednesday afternoon we could sell feeder pigs at the local sale barn. There were other nearby auctions that would take hogs on Monday afternoon, Tuesday afternoon, or Thursday noon. There aren't many of those left now, and some even require "reservations" before you can deliver a set of hogs.

The bridge between the producer who has good, naturally reared hogs and the consumer who wants pork from those hogs has largely been swept away. The contract and the confinement house have taken over where the red clover bloomed and the tall corn grew.

And when the bridge goes down, you're left to ford the creek by yourself. The pork producer's wade through troubled waters now is via niche and direct marketing. Hog farmers now are catering to pig roasts, making and selling sausage, selling boutique pork by the half and whole hog, networking to sell branded pork product bred for flavor, supplying small, rather specialized outlets, and simply building markets anew, one butcher hog at a time.

At the very depths of the 8¢ per pound hog market a few years back, a local sale barn was selling heavy butchers at 20¢ a pound. They weren't that good, but they were going through that sale ring one at a time, and butchers for the home freezer were found nowhere else. In the state known for its corn and cured hams, meat hogs are getting to be harder to come by than feed salesmen that take no for an answer.

Not long ago, a couple of the volume producers in our area stopped me in the parking lot of a local farm supply store to inquire about the progress of our local farmers market. To be more precise, they had heard that some pork was being sold there, and they wanted to know how that was working out.

I told them quite simply that some of the folks there were indeed doing quite well. A couple were netting as much as $100 per head with direct pork sales. Almost in unison, they asked where they could sell 500 at a time at such prices.

I know of no such place, and don't believe one will ever exist. Direct markets are built one hog at a time and one transaction at a time. This is no easy task, and in the early going, selling the product can eat up every bit as much time as it takes to produce the hogs. This is not a market for the dump-and-run product and short-term profit mentality that we have come to associate with corporate agriculture.

A 250 pound butcher hog at 60% yield will produce 150 pounds of whole-hog pork sausage. At $2 a pound, it will produce gross sales of $300. Drop to 50% yield and you still have a $250 gross. Fifty head netting $100 each will yield $5000 in the black part of the ledger.

The volume producer now would be happy to net even $5 per head and would have to sell 1,000 animals to net that $5,000. The range producer working with direct marketing to improve income will have to leave behind much of the volume mentality that has so dogged production agriculture in recent years.

Establishing a Regional Market Base

By industrial market standards the individual producer will find markets for only a relatively modest number of animals, even at the high end of his or her numbers. In direct-marketing circles, it is commonly believed that the key to success is access to a potential customer base of at least 50,000 souls within an hour's driving time of the home farm.

If you draw a circle on a map 100 miles across and with the family farm at its center, this is your base market for direct sales of meat hogs. In most parts of the country, that will pretty well assure you of the 50,000 figure cited above. The more urban that base is, however, the more likely you are to find buyers willing to buy at top dollar. You won't sell a lot of butcher hogs at a premium to other hog producers unless you are a very, very good salesman.

Dealing with this broad base, your early endeavors will largely be what are termed "cold calls," or making first-time contacts. You will be reaching out to people who might know you but who have never associated you with a food item they can buy and use straight away. In addition, your goals will be more easily met if you target potential buyers who are apt buy your product for reasons other than the price.

There have been many studies over the years to categorize consumers, and while all of us are somewhat price motivated, certain segments of the consuming public have demonstrated very different motivations for buying certain goods. There are consumers today, for example, who have expressed a willingness to pay up to a 10% to 12% premium for food items that reflect their concerns about issues such as the environment, human health, animal treatment, and support for the family farm.

You aren't going to find these folks sitting on those beat-up old bus seats at the grain elevator or on the stools at the local diner. Contact points for them are many and varied, and perhaps surprising to many, they are often accessible with a very modest outlay of resources. Here are a few helpful ideas.

1. The very best advertising is by word of mouth. The consuming groups mentioned above seem to delight in finding the people who don't merely supply their needs, but also take the time to interact with them a bit. They will then tell their friends about their special buying experience, and from there all kinds of good word can spread.

We have an Amish friend whose family poultry business grew to 10,000 broilers a year, and not a thin dime was spent on advertising. Buyers from over 100 miles away found their way to the door of this Old Order Amish family.

Business cards with your contact information and a simple list of the items you have available can help facilitate the spread of word-of-mouth marketing when put in the hands of satisfied customers.

2. Simple printed handouts can be sown about places that these folks are apt to frequent. A single-page flyer or even simple folded pamphlet can often be worked up on the family computer. It can then be sent out to health food stores, specialty markets, food co-ops, ethnic markets, garden supply stores, and the like.

They need not be elaborate; a simple, rustic look may even add to their value as a selling tool. Briefly document your farming operation, list what you have to sell, and give information on how best to contact you. Borrow from the big marketers and offer some sort of special hook, such as a discount coupon, a price special, or a "package" deal.

For good tips and cost-saving ideas, try looking online, and don't be afraid to ask folks who might have done a bit of small-scale publishing themselves, such as the person who prepares your weekly church bulletin.

3. Simple ads can be placed in nearby urban newspapers, in the suburban journals or advertising fliers that are common around cities, in local and regional special-interest publications (good choices are those with environmental, food, or lifestyle themes), and even in some of the privately published telephone directories.

4. Become your own press agent. If you've recently bought a new boar or participated in some other sort of livestock event, ask the breed group or other sponsoring agency to send news items to your local media. My senior year in high school, we bought a top-selling boar at a spring breeder's auction and got a Duroc breed association story written that was sent to many Missouri papers and resulted in quite a few inquiries and sales.

In a few paragraphs, announce the launch of your business and send this to area newspapers and radio stations. This is the kind of "feel good" news that appeals to local media in their roles as civic and community boosters. And yes, it really is news.

Submit recipes featuring "range pork" to various publications. Our local paper does a yearly cookbook on newsprint that is sent to all its readers, and those get filed away and referred to often.

Take pork dishes to every potluck, church dinner, and carry-in you are invited to.

Last spring, the "St. Louis Post-Dispatch" announced that it would be carrying a listing of area farmers' markets in an upcoming issue. I sent them a letter detailing our local farmers market and the great variety of livestock and poultry products featured there. The result was that the paper sent a reporter and photographer to our market. They put together a front-page feature in their weekly food section, and even published a detailed map showing how to reach us. Not bad for an out-of-pocket investment of 37¢.

Your Story — Your Message

With the medium for getting the message out resolved, the next selling task is to design and assemble that message. We find that tailoring the product presentation to meet the buyers' needs can boost sales, particularly at the local level.

We sell a lot of whole-hog sausage at our local farmers' market. We have it worked up at a USDA-approved, state-inspected slaughter and processing facility. They produce a product that is approved for us to sell within the state of Missouri. At the facility, they place the sausage in 1 or 2 pound sleeve packs (our choice) and then flash freeze them. With the approval marked on the packaging, and by keeping it frozen, we have a product in a handy size that is readily accepted by most consumers.

With simple signage, we tell shoppers that our sausage (1) is truly made from every part of the carcass, including hams and loins, (2) is leaner than most store-bought sausage products (most of them run just 70% lean), (3) is produced without additives and in a humane manner, and (4) is generally from a rare or endangered breed, such as the Mulefoot.

That latter point is often a good way to begin engaging with potential buyers that are passing by. Folks really want to know how you can help to preserve a rare, old breed by eating them. You can then tell them that not every animal is of breeder quality, that there is always a surplus of males, that they were developed for use as a meat animal, and that this is their traditional role and a part of America's agricultural heritage.

With no middleman in the way of the transaction, the producer can gain insight from his or her customer base and go on to produce a number of near-custom and thus high-selling products. If they like the sausage, you may then become their first choice for hams, pork steaks, chops, and more. This can result in sales of bulk packs of a variety of cuts and/or whole or half-hogs for their freezers. A small family may want to buy a smaller animal, another may, for health reasons, want or even need absolute assurance that the animal was produced without additives, and since it is locally processed, the cuts can be shaped to the exact needs of the consumer.

The range hog produces at once gourmet fare and porcine health food. This is the pork your grandpa raised, your grandma cooked, and that would even qualify for Ralph Nader's seal of approval.

Interestingly, our whole-hog sausage sells as well or better to the local folks than to the more distant visitors to our farmers market. I sometimes think it's because they best know hogs and the right way to produce them.

We have sold a fair number of half and whole carcasses and find that interest in these goes up when concerns about the economy increase. This kind of purchase also appeals to those who shop the warehouse stores and outlet malls. Some day, a group of cooperating farmers will rent a kiosk at one of those malls and book orders for pork there on Saturday and deliver it the next week in a simple, refrigerated straight truck.

A hog delivered on the hoof to a local locker plant will not yield the return of a hog marketed directly as processed pork, but it can still be marketed at a comfortable premium. What is the value of a free-range animal delivered to a processing plant at the slaughter weight personally selected by the buyer? Is it 5¢ or 10¢ a pound more than the current butcher hog market? This is not mere pork.

What about the angle of raising it additive free? Being humanely reared? Being absolutely fresh? All of these certainly now have value in the modern marketplace. It is a value that has not yet been fully established and for which there is little in the way of standards or norms.

If they make a 45¢ butcher hog worth 55¢ a pound, they will do a lot to put your operation solidly in the black. If they make the hog worth 70¢ or 75¢ a pound, it still gives the average family a year's worth of very good pork for less than the cost of one night at most professional sporting events. And that is the way we in production agriculture need to begin thinking and acting.

No one expects America's farmers to become a generation of Donald Trumps over night, but we are going to have to get busy at the job of selling what we produce. The key lies in the figures just given. The range producers are not raising pork for the global village; theirs is production for the concerned consumers of the United States with substantial disposable income. The market that truly and fairly rewards good efforts in quality agriculture lives in the big houses on the hill above the global village.

You don't dump product before these people and then ask them, "What will you give me?" They sure don't sell Rolls-Royces that way. The savvy consumer still harbors a very favorable image of independent farmers as more concerned, caring, and dependable suppliers of food than the agricultural corporations. They know this group has had some hard going of late, and they are motivated to go a bit further to get their money back out to the farm.

You are going to have to go out and meet them at least halfway to get your share of those dollars, however. You must now match your investment in production with one every bit as great in the marketing process.

Sit down and draw up a list of the people you have encountered in the last 30 days. The barber, mechanic, the parts counter guy and the implement dealer, the county agent, the banker, shoe store clerk, the preacher: all these people and more eat pork. They also all understand the business concept of quid pro quo. Some would say that most of us can list at least 300 people we know who represent strong potential markets.

Also emerging now are some novel and potentially quite profitable specialty markets for butcher stock. In the south end of our county, a small packer has emerged buying hogs to be processed into a number of specialty products with an Italian bent. A few years back, the farm press was filled with reports of a Hampshire breeder in Kansas who, through a bit of serendipity, entered the pig roast business. A single inquiry grew into a business that has him selling three to four premium butchers each week, plus soliciting his skills as a roast chef.

The Neimann firm in Iowa has yet to meet the demand for Midwestern-produced, outdoor-raised pork for the very lucrative and fast-growing West Coast trade. Many new firms are rising up in their image to bridge the gap between family farm producers and consumers. In Missouri, a group is organizing to produce and market a flavorsome variety of pork from hogs that are at least 50% Tamworth in their breeding.

In fact, after extensive study, it has been determined that a number of pure swine breeds produce pork with exceptional flavor and other eating qualities. Best known perhaps is the Berkshire, which produces the "black pork" that has been so valued in the Asian trade. A premium of some substance was paid for Berkshire and percentage-bred Berkshire pork for quite some time, but that fell away with the declining Asian economy. Marketing efforts still continue around this breed and the Duroc, Chester White, and Tamworth, seeking to repeat the success of the Certified Angus Beef (CAB) program.

The value being placed on such pork isn't based simply on a whim or fad, either. Berkshire pork has been shown to have a color, texture, and even a pH level that is different from other types of pork. The Berks even put on some fat that is comparable to the marbling effects found in prime beef.

Breeders have begun to assert and promote the fact that some can produce pork that tastes better than others. Range production even builds upon this, as these are the breeds—the Duroc, Berk, Tamworth—most often associated with range production.

Rearing hogs outdoors further enhances pork's eating qualities, due to improved muscle tone, the rich and more varied diet the animals receive, and the fact that it is much less likely to have any type of environmentally related taint. Stand nearby when confinement-reared hogs are being sorted and loaded, and the noxious odor alone will have you questioning how the meat from those animals can ever be made edible.

Relatively new on the scene but growing rapidly is what is being termed "the slow food" movement. It is a grassroots consumer movement that has grown up in response to the continual trend to industrialization in the food industry. They see this epitomized in the fast food sector, and hence the term "slow food" for the counter movement.

With group pressure, political clout, and most importantly, their euros and dollars, people here and abroad are demanding something different and better from the food industry.

At farm conferences, I have been approached by representatives of consumer groups wanting to buy food items outside of normal channels, directly from farmers and growers. A few years ago, I fielded a call from a gentleman in Manhattan seeking input on a project to re-create and market the hams he remembered from his youth. The Smithfield name is very much alive and is a force to be reckoned with in the swine industry, but the Smithfield Virginia ham of old is little more than a memory now.

What the slow food folks are clearly saying is that great change is in the wind. The people that agri-biz believed would consume anything with the right promotion and lowball pricing are saying no. And they are not just saying no, but are going well out of their way to find alternatives.

The outdoor-produced animal is the meat producer that these people and many others want. Those vested in pork as a commodity see no value beyond volume and monotonous consistency within that mass of product. If it meant settling for loin-eyes the size of silver dollars to get them all alike, many folks believe they'd opt for that in a New York minute.

Pork as a product and pork as the traditional meat staple in the American diet diverged quite some time ago. Outdoor-produced pork is the best effort to get them back together.

There are now major firms that are making it work, networks of farmers succeeding with it, and even independent farmers that are carving out niches for the direct marketing of meat animals and pork products. You can't trade in range-fed futures, and there are still a lot of Extension agents who will look at you like you've raised a question about dinosaur farming when you mention raising hogs out-of-doors. The outside hog, the dirt hog, is back. They're beginning to find ready buyers, and this time they are here to stay.

The future of production agriculture in the developed nations of the world has its exemplar not in the industrialized farms of the moment, but the vineyards just now emerging. They produce a premium product for a premium price. It is not volume that keeps them afloat, but rather the creation of high-value products that are highly valued and esteemed by others.

Pork as a mere commodity has failed a great many people, producers and consumers alike. The outdoor producer has ramped up his or her production to something both richer and more unique. It is the good stuff once again.

Feeder Pig Marketing

The range producer has in his or her control something of a secondary product that was once of very great value and that is beginning to emerge once again. And that's the "go anywhere" feeder pig.

If necessary, this animal can go into confinement or onto a feeding floor. It can glean in grain fields following harvest. It will run the woods with the best of them, and it will make money the old-fashioned way, the root hog or die way.

The feeder pig from a controlled environment comes with more conditions and limitations than your great-aunt's Persian cat. There is a 10 pound weight range, between 60 and 70 pounds, where it is believed that the confinement pig can make the transition from a controlled environment to life in simpler quarters. Scrub that if you live where it is cold in January, wet in April, or hot in August.

After we sorted a set of pigs for boars with breeding potential and keeper gilts, we often sold the rest as feeder pigs. There were purebred Durocs, Chesters, and Spots, the supposed "hard sells" as feeder pigs. Coming off of a wooded hillside in northern Missouri, as ours did, they could tread concrete or creek gravel and live and grow in either.

The days when feeder pigs by the thousands moved out of Missouri and other states in the upper South are now past. Southern Missouri no longer proclaims itself the feeder pig capital of the world. There was a time when I could sit in a sale barn where it took hours to work through weekly runs of local pigs that could run to 2,000 or 3,000 in number. They would leave those barns by the truckload for the grain-rich states of Iowa and Illinois.

A feeder pig market of sorts still exists, but it is for 10 pound early-weaned pigs moving from one hothouse environment to another. Missouri-farrowed, early-weaned pigs now may move three counties or three states away, supposedly as a health management practice. It is also a bit of a ruse to work around environmental regulations. Pigs 10 days old or less do not count against the number limits now set for swine farms by regulating agencies.

Feeder pigs were always a volatile commodity, with a demand driven both by butcher hog prices and corn supplies. When rising hog prices and big corn crops combine, feeder pig prices could rapidly spiral to well over $1 a pound for pigs up to 50 pounds. The best I can recall was $1.60 a pound for some 35 pounders. I've sold a few $5 and $10 pigs, too.

What has emerged now is something quite interesting, although not yet fully appreciated or understood. The droves of pigs that 10 years ago set prices for the feeder pig market are no longer seen. They now move almost entirely between contract producers, and there is simply no way to conduct price discovery on these transactions.

The pigs moving in open markets now are generally presented there in quite small lots, and prices very often bear little relationship to either butcher hog or corn prices.

For example, the other day, butcher hogs were under 40¢ a pound, and a caller on a local radio swap show was offering 25 to 30 pound pigs for $50 each. Now, an old rule of thumb held that a 40 pound feeder pig had the same value as 100 pounds of butcher hog on the current market. The producer had that much sweat, feed, and facility equity in the pig by that weight. Another formula held that by the pound, feeder pigs were fairly worth 2 to 2½ times the going price for a pound of butcher hog.

What then makes a 25 pound pig worth $2 a pound in uncertain times?

Well, the corn crop then being harvested had been reported to be the second largest in history, and often the most profitable way to market a big crop of corn is to walk it to town on four good legs. Those farmers wanting a few hogs to feed out for family and friends have been all but shut out of the feeder pig trade. A lot of small-scale and part-time farmers in our area have developed a good sideline venture selling butcher hogs by the whole and half carcass to those they work with and to family members.

Pigs in smaller groups have always sold for higher prices, because of the extra handling and the resultant loss in value to the remainder of the group that had been sorted through and thus reduced in number. I once took a group of about 10 or a dozen Duroc feeder pigs to a local auction where the auctioneer stopped the sale to ask if I would allow the small bunch to be broken apart. The gilts then sold out of the group for $10 to $15 a head more than they would have brought as feeders.

The feeder pig market thus now lies with those who prize and need pigs that have the versatile nature of the outdoor-reared pig. When confinement-produced pigs do appear in general markets, they are severely discounted, due to their small size and the fact that they have undergone major stress in the marketing process. A time or two when prices have really tanked, I have seen these types of pigs actually offered free for the taking.

A well-grown and hardy, outdoor-reared pig in the 40 to 60 pound weight range is a very valuable item. Pack some meat type and lean quality into the mating, offer them in a ready-to-go state, and these days you can pretty much put your price on them.

Making the Most of the Presentation

Not enough can be said for the need to present feeder pigs in a manner that is visually appealing to potential buyers. Every farm is going to get caught with a small pig or two, or even entire litters farrowed out of sync. And there is no such thing as a drove of pigs large enough to hide those odd little guys.

Sometimes you just have to bite the bullet and put them out there all by themselves. Pricewise, they will hurt you less that way than if they were allowed to detract from the value of a large set of otherwise good pigs. I've seen sale rings crowded with droves of 100 plus head of pigs, and the one or two "rats" in the bunch will either get crowded to the outside or even rooted to the top of the group, for all to see. Every once in a while, you will get one that's just so sad it will wander around the place unsold, seemingly for years.

Some of these became legendary, passing through the sale barns of the Midwest in careers that go on for years. To the north of us years ago was a sale barn that hosted a smallish white shoat so frequently that he came to be affectionately known as Lil' Joe.

The return on 10 head of evenly sorted 50 pounders will nearly always be greater than the returns on 15 head of mixed 45 and 50 pounders. They may all be good pigs, but they create the perception of imbalance and potentially uneven performance. The industry is at a point now where some producers are even opting to feed gilts and barrow pigs separately. Buyers are sensitive to mixed sizes, rough and patchy hides, and even a mix of colors. The thinking is that not a lot of planning went into the mating that produced a drove of pigs with all of the colors of the porcine rainbow.

A pig that's ready to go will be:

1. Castrated and long healed, if male;
2. free of internal and external parasites;
3. of a good size for its age;
4. sound on all four corners and is free of blemishes;
5. clean, with a shiny haircoat;
6. part of a uniformly sorted group, according to both weight and quality;
7. offered at a weight appropriate for the season of the year.

Old rules of thumb with respect to the season are that you can sell 'em lighter in weight in the summertime, and you need to buy 'em on the heavier side in the winter. Most producers like to see a fairly heavy dash of colored breeding in their feeders because of the potential meat type and hardiness it represents. Some old-timers believe that these animals cannot be too Black or Red when the snow begins to fly.

The outdoor feeder-pig producer generally operates with a three-breed crossbreeding rotation to maintain optimal hybrid vigor in the pigs that are going to be sold. Here is where the old Black/White/Red rotation really made its name. The sandy Red shoat has had its supporters for about as long as corn has grown in Iowa.

Our purebred feeder pigs always sold well, but the first-time sale to a new buyer sometimes took a bit of salesmanship. The purebred hog still suffers from a bit of a "tender" image with some folks. We may have actually coined the "try 'em you'll like 'em" sales campaign Madison Avenue later put to use.

The good feeder pig doesn't need a lot of selling if it has been bred and readied for its life in the finishing pen. Until just a few years ago, it did need to be offered in groups of at least 30 to 40 head (a minimum of three to five litters) to achieve top dollar. The large draft market has now largely been gobbled up by the corporations, and the premium now falls to small numbers that need minimal care to get off to a good start in life.

One, two, three, . . . anything less than a dozen or so feeders are destined for special purposes and niche markets; they will neither fit nor find their way onto any market chart or reporting service. They are worth whatever the buyer has to pay if he or she is on the receiving end of that niche business. Thus, feeder pigs can be a very valuable commodity even when the general butcher hog market is in a steep downturn.

If such buyers are not at hand, however, then pricing can become a very precarious thing indeed. Recently, in the space of a scant few months, I saw the price of 40 pound pigs at auction swing up and down by nearly a full dollar per pound. It can truthfully be said that the feeder pig market has lost none of its volatility even as it has moved downward in scale.

In fact, I seriously doubt that it ever will. Even back in the day, I shared a belief held by many others that it was just too great a risk to become a feeder pig production specialist. To us, feeder pigs were always a secondary market, sometimes even a fallback if we needed it. The money has often been quite good, but you must always be mindful of what is driving that market. I am thinking particularly of the few times when the market was hard driven by the limited availability of pigs. The few times we saw 40 pounders crowding $60 per head, that also meant that butcher hogs were pulling in $150 a head.

Feeder pig sales make good sense if you are anticipating a falling market, or else if factors that drive the market go beyond simple supply and demand. Sometimes in bumper-crop years, feeder stock has been driven well beyond reasonable norms, because they are the only other marketing option for coarse grains. That's a kicker that remains in place even if the numbers driving those regular feeder pig markets drop significantly. In the fall of 2003 in our area, many folks were searching for pigs ahead of a big local crop harvest, as feeder cattle prices moved quickly to record highs.

We have sold feeder pigs when we had space problems or when they were worth more to others than to us. They are a good alternative product for a swine venture and can often be counted on to provide some early-stage cash flow. However, the risks of having feeders form the sole product of an operation, or else limiting your marketing options to this one aspect of the venture, are just too great in most instances. Feeder pigs are best exploited when used as an additional marketing option, to lend diversity to a swine enterprise.

Seedstock Marketing

The seedstock arena has always been a somewhat heady venture, but historically it has also been one of the bulwarks of the family farming tradition. All across the nation, there are multigenerational farms producing seedstock, as well as state fairs regularly honoring farm families who have been participating for some 50 years and more. Many of the pure swine breeds still have third- and fourth-generation producers in place.

The role of the seedstock producer has been discussed and debated to quite an extent over the years, most folks believing that the breeders should occupy no more than 5% to 10% of the total producer population. These are the producers who develop breeding strains, shape animal types to meet demand, and offer animals of good breed and economic type to the community of general producers.

This is not to say that there isn't enough room or that there isn't a need for new and dedicated breeders. We once lived on a county road that had a purebred swine producer in place for nearly every mile of that road. Four purebreds and a great number of gilt crosses were available along that little stretch of blacktop. And many potential buyers with different price ranges and needs were drawn along its bends and curves.

When it comes to pricing and other business practices, the seedstock sector is certainly going through some trying times. A handful of boars recently sold in the low six-figure range. They have since gone on to perform as artificial insemination studs, from which two sticks of semen and a breeding certificate may set you back a cool grand. No doubt, the "boar in a bottle" has killed a lot of boar sales. The majority of breeding-stock sales these days are compressed into a fairly tight range of just a few hundred dollars per head.

It's not bad money, but where once a small producer might have sold 40 to 50 animals a year in that price range, now it may be more

A Few Words of Wisdom

"The problem isn't that small farmers aren't economically efficient. It's that industrialization leads to closed markets where prices are fixed not by open, competitive bidding, but by negotiated contracts."

"The good news is that there are cost-effective, efficient ways to produce hogs outside of total confinement—whether it be in lower cost facilities or on pasture—that can compete on cost terms and respond more directly to the social, environmental, and health concerns consumers are interested in." Chuck Hassebrook, Center for Rural Affairs, Walthill, Nebraska

like eight to 15 young boars a year. Still, as long as it remains in the hands of independent producers, purebred seedstock production assures the family farmer control of the most important of all production resources—the seed, the basis of all production. With free access to seedstock and the markets, the independent producer will nearly always find a way to stay in place and to compete. Remove either one or both (as some have tried) and the game is finished.

You can be sure I am not a big fan of artificial insemination. We have it here and have to make some kind of peace with it. Still, we must never become too comfortable with it.

Seedstock production was once regarded by many producers as the province of the wealthy and the big operators, but it was never really so; it is a resource that's simply too valuable to have been left to such few breeders. We actually come from a tradition of yeoman farmers who kept bloodline horses, well-bred dogs, and gamecocks with centuries of breeding behind them, and who had an eye for good stock that began with the sons of Adam, the first herdsmen.

Sadly, we have come to a time when some breeds of swine are in as much danger of extinction as many creatures in the wild. And this doesn't just include the nearly feral and always minor breeds like the Guinea hog and the Ossabaw. Historically important breeds like the Hereford, Tamworth, and Mulefoot are now quite rare; the Mulefoot was down to its last three breeders just a very few years ago.

As an aside, one of those fellows, an older gentleman, lives not very far from us. He was a regular at the farm auctions our family business held on occasion. I've written about the old farmer and his hogs in a couple of articles. He's thanked me for it and the boost for his hogs, but has let me know in no uncertain terms that if one more socially aware, environmentally sensitive sort from the Northeast woke him up at the crack of dawn for a little chitchat about the philosophical nuances of the oneness of nature, he was going to hunt me down and unite me with my farming ancestors.

Now even once-common breeds like the Chester White, Black Poland, Berkshire, and Spotted are beginning to slip away at a rate that should have everyone concerned. A handful of minor and fast-disappearing breeds that should be of special concern to the range producer are also the Tamworth, Gloucester Old Spot, and the English Large Black. These were the original "grazing" breeds, and while they may be a bit behind the times for some type qualities, they represent breeding resources that may yet have their best days before them if taken up by dedicated producers.

Fortunately, small but dedicated numbers of folks are strong advocates for most of the breeds mentioned above, many of which still exist in large enough numbers that they are considered genetically strong and quite viable. The market they are currently serving is made up of other breeders, farmers producing butcher and feeder stock, and the rapidly growing show-pig trade.

Boar-Buying Basics

Once upon a time, the basic breeding-stock sales unit was the breeding-age boar, and they were sold by the thousands each spring and fall.

Pegged at being worth 2½ to 3 times the value of a 250 pound butcher hog, purebred boars of 7 to 10 months of age sold readily. Health security measures began tightening up about 30 years ago, and this rather effectively shut off the trade in auction-house or used boars. It resulted in roughly a 2 decade long run in which the young purebred boar bought from its breeder set the market and led the industry.

When the big corporations began to form, the purebred sector deluded itself into believing it would have a continued role as the breeding stock supplier to this 800 pound gorilla in their midst. Too late they learned that that particular gorilla was going to function only with what it could and would produce in-house. This, coupled with the growing role of artificial insemination and the use of composite boars, meant a rapid and rather painful restructuring within and outside of the purebred sector.

The breeding boars were driven out of the seedstock showroom and replaced by the show pigs and the gilts that would produce them. The breeding-age boar didn't exactly become a bargain-basement product, but its role and its value was seriously reshaped.

Hurt especially hard by sharply discounted salvage values, young boars are now primarily marketed to the independent producers in modest numbers. One of our main buyers was once part of a farming partnership of four brothers who would buy 8 to 15 boars from us each year. Now most report boar sales of just one or two head at a time. It is an unsettling time, but one that has inspired some very imaginative marketing techniques.

A lot of producers seem to be borrowing from the automobile trade, offering "$250 buys any boar in the lot" promotions. Others are taking up modest numbers of several breeds to position themselves as a sort of one-stop shop for boar buyers. A friend in western Illinois now has Berkshires, Herefords, and Tamworths on his small farm.

One marketing "gimmick" of sorts that caught my eye a few years ago was simple yet very intuitive. The producer continued to raise three breeds of boars in decent numbers, in addition to row crops on a substantial scale. One of the consequences was that he didn't have the time he needed for the one-to-one marketing of his animals. So he created a method of presentation that could be used by any family member or farm employee available to present the boars when prospective buyers arrived.

As the young boars were developed and handled, he would sort them based on visual appraisals and performance. The very best would receive a purple ear tag, the second best would get blue ear tags, and the third-rung pigs got red tags.

You've probably already noted that these colors correspond to the color of the show ribbons presented to the winning animals at county and state fairs. The purple ribbon went to the grand champion, blue for class winners, and red for those that competed well. It's an association he expects his potential buyers to make as well, and he's priced each group accordingly.

A young boar should easily service 10 females if fairly matched for size. If he is being hand bred or well supervised, a few more may be possible. Producers with boars to sell have firmly touted that ratio over the years.

Most young boars come with a warranty of sorts that is based on the terms of their use. Typically, to guarantee that a boar is sound, fertile, and serviceable, he must meet certain conditions: (1) he cannot be ringed, (2) he must be hand bred in the early services, (3) he must be kept well bedded and properly tended, (4) he must not be worked with other boars unless he was raised with them, and (5) he must not be put into service until he is at least 8 months old.

We have had to replace very few boars over the years in order to make good on this guarantee, and I can honestly say that most were due to buyer neglect or abuse. Sometimes, it is easier, not to mention better, just to pull the animal out of a bad situation. We once sold a set of 5-month-old boars to a fellow who sat through our simple list of do's and don'ts, nodding all of the way along. When we delivered the animals, he had us unload the pair of youngsters into a pen of over 50 gilts of various ages. He didn't like it when we told him all guarantees were off, but he liked it even less when Dad said we'd put them back on the truck if that didn't suit him.

Our goal has been to produce an average of 2 to 2½ marketable boars per litter. Our best was five boars from a couple of litters. Our worst was, well, let me just say that we have made our fair share of contributions to the cull sow trade.

Picking boars to sell begins the day they are born, and most good ones weigh at least 3 pounds at birth. The best pig we ever raised weighed a bit over 4½ pounds at birth and stood out throughout his entire time on the farm. If marketing to other small and independent farmers, and especially other range producers, you should avoid selecting for extremes in type, however.

Granted, animals that are extreme in some areas are needed from time to time, but these are boars that are there for a special purpose and therefore must be placed carefully. Outside of those particular parameters, however, they can be a problem.

We once had a very demanding neighbor come over to buy a pair of Chester White boars, and I knew we were in trouble when he selected what was perhaps the longest and leanest Chester boar we ever produced. That animal was probably too lean for most breeding needs, and I knew that in that producer's pens, the other heavier boar would work rings around him. The pigs he did sire, however, were clearly his—long and lean.

My neighbor found plenty of opportunities to comment on that old pig's rather spotty service record, and many times I offered to buy that pig back. But that pig hung over me for a couple of years, through no real fault of his or my own. In a herd with a real fat problem, he would have produced a set of replacement gilts that cleaned up substantially but still had good cover. In situations like these, sometimes the hogs' faults fight it out with its strengths, but without a clear winner. While I may come across an animal like that one again, I'm going to be mighty, mighty picky about who gets him.

Boar sales are gradually coming back, and many small- and medium-sized producers are now having to scramble to find boars for natural service operations. Artificial insemination has considerable hurdles to overcome before it is widely accepted: is far from foolproof, it can be quite costly, and there is substantial resistance to this technique among end consumers. It may not even be allowed in an operation seeking organic status. The buyers now in place will need replacement boars at roughly 18 to 24 month intervals, unless you are working with multiple breeds and can feed into their crossbreeding rotations more often with unrelated animals.

A lot of the traditional boar marketing outlets are basically history: the spring and fall breeder sales are nearly all gone; few now are the seedstock auctions of champions following state fair shows; breed conferences and other such national events are in marked decline; and very little remains in the way of university-sanctioned boar testing and sales. That said, it should also be noted that boars at the very top end of the spectrum are selling for prices that would never have been dreamed of when all of these institutions were commonplace.

The Feminine Side

The other half of the seedstock trade, the bigger half right now, is the production of bred and open gilts. The thinking long ago in the seedstock business was that while the boar sales made the headlines, it was the gilt sales that paid the bills.

Purebred, F1, and even some F2 gilts were and still are big business for a number of producers. They're often moved about in lots of 100 or more head to producers who are repopulating or creating new ventures.

Reflections on First Ventures in Hog-Buying

For most of the fellows of my generation, the first big purchase of our lives was a gilt. On a chilly autumn day in a drafty old barn, we would sit pressed down between Dad and a 4-H or FFA leader, nervously gripping a sale catalog rolled tightly in our hands.

Seated around the ring would be a dozen or so youngsters like ourselves, knots of older boys in FFA blues, farmers from near and far, and breeders of every tier. Some would have national reputations, and we followed their achievements in much the same way other youngsters followed athletes or performers.

Looking back now, I believe we beginners knew and worked the gilt market about as well as anyone ever has. We each sat there with X amount of dollars to buy Y amount of gilt. You could tell when dollar limits were being pushed by Dad's grimaces and long looks. The first lesson of a youth project is generally one of the economic realities of the business.

Most of us became clear-eyed little capitalists quite quickly, however. We would study sale catalogs for weeks ahead of the big day, narrow our potential choices to those we had a realistic shot at, and then begin weighing carefully how we could go about making their offspring better. I can close my eyes easily now and imagine a knot of us working through a pen of gilts, looking for the one that would move us a little farther down the road without breaking the bank. When you're buying them with hay-bucking and farm-chore money, you have the aches and calluses that come along with the dimes and dollars, and you mean to make them go as far as you can.

You don't go looking for the best, but you know all of the pens where her sisters can be found. You look among the youngest in the offering, as they are the most likely to be discounted, and you are willing to live with the blemish or the fault that some others might not.

I go to hog events now hoping as much or more to see such eager boys and girls again as to view good stock and visit with old friends.

Sales in those numbers aren't often seen outside of contract production. The gilt as a youth-project animal is the dominant market shaper of the moment, but sales in a variety of other outlets remain strong. In fact, gilt prices have shown very little real decline, and they are now the featured item in the breeder sales that remain.

Traditionally, gilts from the White breeds and their crosses sold better than colored-breed gilts. They were, after all, being bought first and foremost to produce pigs. Hampshire X Yorkshire gilts were real treasures, selling by the head in the price range of good young boars, often bringing in far more than all but the choicest purebred gilts.

Young gilts are generally sold in the 6 to 9 month age range, close to breeding, and weighing around 300 pounds. They are offered up well grown, free of parasites, and ready to slide into the breeding herd. Good gilts nearly always seem to be produced in greater numbers than good boars, but that doesn't mean they can be taken for granted.

The purebred gilt has been unjustly feared for her effect on a good commercial herd. I don't like to see in any herd a motley array of colors and types, and if they have been put together right, purebred gilts will mother up and raise pigs in decent numbers. For example, we pushed hard in selecting for both body length and litter size with our Durocs, and our Red gilts "nicked" well (effectively blending the various aspects), whether bred to Black boars, White boars, or Red ones.

Still, the range producer is perhaps more likely to be producing good crossbred gilts for sale to other independent producers. They will generally be a Black X White F1 or a first-generation cross of two pure breeds or a Black X White X Red F2. This will be a Black X White bred to a pure Red breed to produce a desired cross deep with hybrid vigor. The Hampshire X Chester White gilt has always been appreciated when she has been available, which is not nearly often enough.

Currently, a set of good Berkshire X Chester White gilts should be a good mix. These two breeds are beginning to carve out a niche for the quality and flavor of their pork. Such a cross might be a good sideline for herds of one or both of those breeds that are kept to produce purebred stock.

Through the years, I have seen several attempts to create a "range" gilt to be marketed to folks producing butcher and feeder stock out-of-doors. The breeds tapped for such a genetic mix included the Tamworth, Hampshire, Yorkshire, Black Poland, and Duroc, among others. The more genetic pieces you try to combine in such a mix, the less predictable the outcome becomes. I'm not saying this is an idea without merit, but it must be started with a very clear plan, be kept as simple as possible, and may best be achieved by two or three cooperating farmers working on separate farms.

At the risk of sounding a little simplistic, I have to say that good gilts are worth whatever they will bring. I have never seen anything even remotely akin to a dependable pricing formula for open, ready-to-breed gilts. For many years, a personal parameter of mine was $150

The Allure of Purebreds

The purebred producer must be very careful about not getting too hung up on the phenomenon of selling-price inflation. Every show season produces a handful of hogs that sell for Las Vegas prices.

They get written about like rock-and-roll divas, but just how important are they, really? Veteran producers know these big show pigs are seldom heard from again at that level. It does often seem that they make better grandsires and granddams than they are producers of winning sons and daughters.

Sure, the earning potential of a good gilt is greater in most instances than for a hog readied simply as a butcher. Still, the market for her will be rather finite. It is a market in which it takes time to become established, and there is a bit of a selling job that comes with them.

The successful seedstock producer doesn't let himself or herself be too narrowly restricted by that title. For 30 plus years, I thought of myself first and foremost as a producer of purebred Duroc, Chester White, and Spotted boars. In total, that represents well over 100 pig crops, and included in them were a fair number with little or nothing in the way of keeper boars. Other times we wound up butchering and eating some right good boars because no other profitable outlet for them was at hand.

In some crops, our main profit was from feeder pigs. From others we kept and sold a few more young boars than perhaps we really should have. When hogs crack that 50¢ barrier, you can sell just about anything with a snout and a corkscrew-shaped tail. When butchers start dipping below 30¢, you will have trouble drawing buyers with even a two-for-one sale. I have seen that tried once, too.

Still, the purebred herd gives the outdoor producer perhaps his or her greatest number of marketing options. Good breed type and basic, real-world characteristics can be fitted into one animal. And more likely than ever now, you will be selling a bit of everything from each pig crop farrowed.

to $200 per head, or roughly $50 to $75 a head over what they would sell for as butcher stock.

A lot of folks buying gilts in real volume would fight hard for a standard at $25 per head over butcher price. It could be done if they would accept "gate cut," taking a certain number of animals straight out of the pen.

As we began to tighten things up in our operation genetically, we moved into a bit of a different gilt pricing zone. I once bid to $400 on an open gilt, and decades later I can still argue both sides of going the extra $100 more that it took to get her bought. That is still approaching top-end money on gilts, although you may hear of one now and again that gets into the low four figures. We have found a lot of herd builders in the $180 to $250 per head range, and that still seems to be holding up rather well.

Bred gilt marketing has a second option that has seen a real resurgence. December and January often see some producers collecting a nice second check from their spring pig crop. It is the season for selling bred gilts for spring litters. This used to be the second sale season, a time when you would see a lot of late-farrowed April, May, and June boars and gilts, put up for auction.

A good bred gilt is worth at least $75 to $150 more than her open sister. She will have been on the farm of origin several months longer, eaten a good deal more, and tied up extra facilities. And then there are often the added costs needed to get them bred and clearly showing that they are "piggy" (*i.e.*, with pigs). They should be bred to boars of good quality.

With the market for show pigs now driving the purebred market so strongly, the bred gilt has to be considered its Cadillac product. A lot of them sell one or two at a time to 4-H and FFA youth. Where good ones are available, they represent the very best way into swine production for just about anyone.

Bred gilts certainly aren't doily delicate, but they are still best moved from farm to farm either in the first few days following breeding or within a very short time of farrowing.

Most are offered up close to farrowing, when it is obvious that they are packing their own guarantees as bred females. Gilts bred for January, February, and March litters are especially valuable as producers for youth-project work. Their offspring will be the right size for the county and state fairs later in the year.

The Show Pig

You didn't get to see it, but I took in one mighty long breath before writing this. I showed hogs when I was a youngster, and for 9 years Phyllis and I led a 4-H club and worked with youngsters readying market hogs for one of the most competitive county fairs in the Midwest.

A top-placing pig may sell for $5 per pound in the sale of animals following the shows at our Lincoln County Fair. Winners at national shows may sell for much more, and college educations can be funded in good part with the money from showing hogs and other livestock.

The show pig is a young shoat that is grown and fitted for market hog competition at a fair or other competitive exhibition. Sure, it's a feeder pig, but it's oh so much more. The show pig is the porcine equivalent of the club calf, but it has been turned out with a few unique twists of its own.

A 4-H or FFA youngster will generally purchase from one to five of these young, 30 to 60 pound pigs to grow and fit out for the market hog shows in his or her area. A number of purebred folks and a few pig jockeys are getting into this, and a fair number of jackpotted local and regional pig shows are even beginning to emerge.

This can be a pretty heady game, and there is some very real money on the table. A couple of local kids traveled across two states and laid down $10 a pound for a set of pigs awhile back. Some 140-odd kids weighed in more than 300 pigs for our last fair. Multiply that by the other 100-plus counties in the state with agricultural-based fairs and you quickly get a feel for the potential of this market.

States as diverse as Texas and California are strong seats of youth showmanship. Spring seedstock sales have been replaced with spring show-pig auctions.

The modern show pig is very, very extreme in type and leanness. They can be almost exaggerated in their appearance and even somewhat exotic in their genetic composition. They are porcine hotrods, if you will, and can be subject to a fair bit of tweaking along their way to the show ring. Some now may even have a heavy dollop of Pietrain breeding, picking up a bit of that distinctive, double-muscled swine breed from Europe.

A few swine-show judges are trying to keep at least one foot in the real world. You now even hear stories of judges turning some hogs away from the show ring simply for being too extreme in type to be practical. Very heavy muscling can lead to soundness problems. Most of these very heavily muscled animals are poor candidates for the breeding herd, as those who have bought market-show breeding gilts have discovered. The mating that produces a market show pig these days is, for all practical purposes, considered a terminal mating.

This is because a good show pig is bred essentially for that one day in its life when it passes before the judge in the show ring. Its genetic composition is skewed heavily for muscle definition, and it has been fitted to an extreme. There are now all manner of special show feeds and ration boosters for pigs being readied for exhibition.

Fat cover has been pared unmercifully on these animals, which now strongly resemble jeweled timepieces in their structure. At auctions, such pigs are often presented on display tables, much like ladies' handbags and hats.

I realize I may come across sounding rather negative about this market, but this is not really intentional. Phyllis and I were in 4-H work for years, and youth animal projects are excellent teaching tools and character builders. It is a trade that has some real characters in it, however.

These animals provide an especially good project for younger folks who have an interest in showing animals. They are fairly docile and not nearly as daunting in size as a steer. It amounts to a youth project that has modest costs, simple space and facility requirements, a fairly short time span to project completion, and there is quite a bit that the youngster can learn to do to improve the animal's prospects inside the show ring. To be sure, you can buy a show winner if you're so inclined, but a lot of really good animals have also been messed up with poor care and pitiable showmanship.

At most shows, the youngsters are allowed to weigh in three to five pigs at the preliminary a few months ahead of the actual show. There, they are generally permitted to show from one to three pigs. At most prelims, the pigs have to weigh between 35 and 75 pounds in order to be considered a good butcher weight for the actual show that follows 12 to 16 weeks later. Pigs for market hog shows are normally required to weigh between 220 and 300 pounds. Prime competing weight is 240 to 260 pounds, with the hottest competition generally in the 240 to 245 pound class.

More and more, performance testing is becoming a factor in market hog shows and the ranking of the animals. At our fair, all the pigs are ear-tagged with individual numbers at weigh-in and are automatically entered into a rate-of-gain competition. There are awards for the fastest-growing animals, but these are often not the trimmest and fittest animals in the show ring. Scanning devices now make it possible to gather loin-eye size and back-fat measurements at ringside, and this data can be used to determine class placing. And some market hog shows are followed by a carcass show, where the hogs are slaughtered and then judged on the rail for carcass merit.

Most market hog shows are "terminal" shows: the animals go direct from the show barn to market. They are targeted for that one moment in the spotlight of the show ring. Gilt pigs will grow slower than barrows, but they tend to hang a leaner carcass if a rail competition is part of the show. Barrows will grow faster and carry a bit more finish, but they also tend to be a bit larger framed and may hang a longer carcass.

Dark-colored hogs show better under the lights at night, and white animals tend to stand out a bit better in daylight shows.

A successful producer in this market knows the shows in his or her area, and maybe even more importantly, knows the judges and their type preferences. Here in Missouri, a youth livestock judging competition and field day is held early each spring at the University of Missouri. A number of livestock judges and youth advisors attend, and it's a good place to show up to get a feel for what the judges will be looking for once the show season begins. Similar events are held throughout the country, and your local Extension agent should be able to advise you about the dates and locations.

I like to see a youngster start with two or three pigs, since these are social animals and because accidents can happen. My first show barrow injured an eye on a tie wire end not 3 inches long. It went into shock and died within minutes. Having several hogs may also give the youngster the opportunity to participate in more than one weight class or show.

After 2 or 3 years of competition, I like to see a young person make the next step up and begin raising his or her own show pigs. This is not always possible, and it does increase space considerations and commitment considerably, but it is the best way to add to the young person's learning and life experiences.

At our local fair, the older youth help the younger and smaller ones, and their maturity and added size are much appreciated by all hands in the show ring. However, they tend to have an edge that comes with experience and refined showmanship skills. For this reason, I would like to see more youth shows further divide the age divisions of the exhibitors. A few shows do manage to break the showmanship classes into age groups and evaluate the youngsters' poise and skills as they show their animals.

The range producer considering this market will no doubt be drawn to it by the selling price for these pigs, but keep in mind that this is a very exacting venture, with seemingly ever-changing type parameters. Show pigs are generally farrowed in the very coldest or hottest months of the year, and this can be one of the greatest challenges to outdoor producers with simple facilities and a more seasonal system of production.

As for the question of extreme type, it appears that this is starting to normalize somewhat from recent extremes. You could liken this trend to the "baby beef" fad that swept the beef industry shortly after

World War II. At that time, beef producers fixated on an animal with an economic value based on artifice, rather than natural merit and real-world utility.

That fad passed quite quickly after a run of a very few years, and beef animals were once again bred to more practical norms. The good range-developed hog has the frame and vigor to carry good muscling and growth, and also to carry it all the way to the rail.

And on that note, I have to say that range producers have more than just good hogs to raise and bring to the young people that make up the greatest part of this market. They are producers who are not only working with the land, but who are incorporating a real system of values and goals into their work. If we don't make an effort to practice these ideals and share them with others, the next generation of producers will see this as another market on to which they can dump a few high-dollar pigs and then move onto the next big thing.

The good range producer is not a pig jockey. He or she is selling a message with the animals. And it is a message that the producers of tomorrow particularly need to know.

Purebred producers must function with a clear view of the present and a sensible vision of the future. They are supposed to be doing what they do both for the good of the breed and the good of the producers who will be using their hogs.

What the range producer with purebred genetics is doing right now is assuring a future for swine production, whatever course it may follow. Confinement production has no such flexibility and no other options. It is a long, dark, one-way road. The hog that's fitted to the gestation crate and the 8 square feet of a confinement pen is not going to fit anywhere else.

What Else?

Can anything else be done with the range-produced hog? Well, yes, but here is where the producer's imagination and creativity comes into play.

Hogs are not exactly companion animals, nor are they draft stock. (How would you keep a yoke on a hog's neck, what little there is of it?) Yet they do have some novel uses. Old-timers believed that the wallowing activities of a few head of hogs could sometimes seal up leaks in small ponds. It was a sort of pig-as-putty-knife theory, but it also had its downsides, which is that they can do considerable damage to dams and banks with their sharp hooves and rooting abilities.

I have seen hogs put on brush-clearing jobs similar to the ones normally set aside for goats. Even a well-ringed hog will flip up the dirt in mud season, and they can take a toll on good timber but will clean up low brush and scrub quite quickly. We have used sows to knock down fescue ahead of the plow, too. In one small pasture of mine, a set of sows threw their rings and rooted some fescue hummocks up into small mountains. Talk about fun to plow.

Many outdoor producers have become quite active in swine breed preservation work and have found a rather unique production and marketing niche with these old breeds. They are the porcine have-to-have for the new "rurban" set, and the scarcity of supply certainly adds to demand and selling price. Breeds like the Mulefoot were historically very important, and they are featured now in many historical and agritourism spots. Groups like the American Livestock Breeds Conservancy have even found film and TV work for some of these great old breeds.

The Guinea hog and the Ossabaw are essentially feral animals, and some folks have raised questions as to the absolute purity of their breeding. Likewise, the presence of wattles or waddles doesn't make every hog a Red Wattle. Yet these animals represent a genetic resource that has sparked much interest of late. I'll concede that extinction can be a natural process, but at this point in time, I don't think that those of us in production agriculture can afford to let anything go.

Here in Missouri, we are even seeing some folks working with range-raised Russian Wild Boars. I don't know for how many generations they have to be bred in Missouri before they get to drop the "Russian" tags, but some may be getting close. Some are sold to shooting preserves, but a lot are going into the specialty or exotic meat trade. After all, wild boar is the Saturday-night supper of old British royalty, isn't it?

Actually, the interest in wild boar moves up and down as the use of game is touted in the food press. A few upscale restaurants regularly feature venison and boar, and their meat is nearly all farm raised. Wild boar is viewed as both flavorful and healthful, and it sells accordingly.

From what I gather, they are far more fearsome in image than in actuality, and can be contained with rather basic fencing. They are seasonal breeders and produce rather small litters, but some producers are using them in interesting crosses with swine breeds such as the Tamworth.

Breeds like the Mulefoot, Large English Black, and Wessex are the most distinctive in type and breed character and have exceptionally strong ties to small farms and traditional farming practices here and abroad. They weren't packed up and taken inside a few decades ago, when the big shift to confinement rearing was made. They were left outdoors, through no fault of their own.

Whenever I pass a feeding floor and catch the distinctive characteristics of Landrace breeding in a pen's occupants, I have to wonder why the Wessex option wasn't considered. The two breeds are very similar in type and performance. The Wessex was the preferred breed of many outdoor producers in years gone by, and the belted pigs shown in the illustrations for many children's books were Wessex in body type.

The Wessex and Large English Black are two now rare breeds with long histories as dirt-farm hogs. They are bred with a body type similar to the Landrace, but are hardy and more durable. A very few can still be found in the U.S., but the Wessex once had some real fans among producers all along the Corn Belt.

We have worked with Mulefoot hogs, and in a very few generations have seen them respond quickly to selection for improved meat type and length. The Large English Black is now seldom seen in North America, but it is a big-outline breed with the potential to improve both litter size and carcass length.

We have sold a lot of whole-hog sausage made from surplus Mulefoot males and females that did not make the cut as good breeding stock. The rare breed aspect has in fact proved to be something of a selling tool. If for no other reason, people stop to ask how they can help to preserve a breed by eating them. It can lead to some awkward moments as you try to explain the sex life of the hog to even well-meaning city folks, but they quickly accept that these animals were developed as meat animals and that this is just the last and fitting part of their life. They often get a bit of a kick out of knowing that their grocery money is going to help keep an important aspect of the nation's farming history alive.

I don't know that any state needs hundreds of historical farms, but agritourism is growing by leaps and bounds and is yet to be fully tapped. Historical farms are now emerging to depict specific eras from Colonial times to more recent times, such as the late-19th and early-

20th centuries. Heck, the tractors I grew up on are now showing up in antique tractor pulls. Many of these farms now feature "heritage" livestock breeds, including those mentioned above.

The alternative to today's industrialized agriculture is actually a return to more historically valid and grounded farms. The farms of the '40s and the '50s, when outdoor production was the norm, may have been closer to a farming ideal than many might have imagined at the time. They had a balance and a harmony that we haven't seen since then. They boasted a richer variety of products, and certainly supported a far greater number of farmers. Nothing has worked as well since.

A range operation that has heritage breeds of hogs, as well as other rare and heirloom livestock and poultry varieties, diversified crop plantings, and gardens is a marketing engine. From Jefferson and Washington to Bromfield and Leopold, and on to the health and environmentally concerned folk of today, this system epitomizes the American agricultural ideal. The people want to believe these are the kinds of farms that are feeding them now.

More and more farmers are tapping into this national desire by inviting consumers onto their farms to directly buy the goods that are produced there. In such situations, the producers are paid not just for the actual product, but also for the entertainment factor and their own good efforts as wholesome producers and preservationists.

Profit is where you find it in this day and age, and in the hog you have an animal with a high level of appeal and entertainment value. If you doubt this, take some time at your next visit to the state fair to check out the pig races and the petting zoo or barnyard. People like their porkers, both on the plate and on the hoof. And when they're on your farm, you are the only one they will buy from.

A back-to-the-land movement of sorts has been continuing intermittently at least since the '60s, and many stock raisers have found a rather interesting and often profitable role to play in it. The part-timers, sundowners, weekenders, "rurbans," hobby farmers, or whatever you call them are eager to stock their "farmettes" with at least a wee bit of everything. And along with all of these heads of livestock, they need access to the how-to of caring for them.

At our local farmers market, I have sold a number of hogs to such folks and have been asked to advise them on their care and management. The phone here rings two or three times a week with hog calls from people who have read articles in one publication or another or got my name from a friend. I also field hog-related questions for the Missouri Alternatives Center, and even when hog prices were at their lowest, that phone never completely stopped ringing.

What we have learned as swine producers has value to those who are new to our ranks, and this expertise can be marketed to them. Sometimes it can be added into the selling price of the animals, sometimes with the sale of support goods or with follow-up services.

You might call this farming the urban sprawl, and it is being done more and more where newcomers and veteran farmers have the opportunity to collaborate. I now list consulting as a sideline business on my business cards. A few years ago, a friend from high school opened a novel small business refitting hog houses and other used livestock equipment for resale. He worked from a sort of used-car lot setting and was able to sell equipment, some hogs, and his services as a broker of both goods and information.

We are living in an information age, and the life experiences of veteran farmers and livestock raisers have a very real cash value. There are evenings here when I may spend as much as 3 or 4 hours on the phone. If it is with buyers of my animals, I see it as just another part of market building and servicing. If it is someone wanting to pick my brain, I don't feel uncomfortable asking for a modest hourly fee. Our local mechanics now get $35 an hour to work on our automotive problems, so I feel justified in asking someone $20 to $25 an hour to work on their hog problem.

The Mechanics of Marketing

As noted above, the selling aspect of the business is a very difficult task for many people. We weren't all born to be Fuller Hog Men. It is hard to step forward and take a salesman's role, but it is so very sorely needed now. And anything would be better than the dump-and-run marketing that has been practiced so often in the past by those of us in production agriculture.

There are some simple measures that you can follow that will help greatly in the all-important selling process:

1. Have some really good business cards made for your farming operation. Keep the design simple; don't clutter the card face, but make sure it lists all the things you have to offer. Handing these over to friends and prospective buyers shows that this is a serious business, and more often than not, these will get filed away for later use.

I know farmers who clip their business cards to the receipt of every bill they pay and drop one into every business letter they send. Phyllis won't let me put them in our Christmas cards, but you know I have thought about it.

2. A good letterhead and envelopes can be useful selling tools, as well. Again, keep your message simple. Black ink on a white background is a mark of true professionalism. A simple graphic can be helpful in conveying your message, too.

3. A road sign that is well designed and maintained can be useful both in drawing passersby and pointing them in the right direction.

The lettering should be at least 4 inches high in order to be easily read by passing motorists. Use dark letters on a light-colored background, and keep the message short and simple.

Try to avoid making your signs too cutesy; no sows in dresses or bonnets, for example. A couple of years ago, some friends posted a sign for their rabbitry alongside a busy county road, and soon afterward they suffered the theft of some very valuable breeding animals. While it's hard to say for sure that the two events were related, they had, in fact, recently enjoyed some well-publicized show wins and had erected a very detailed sign.

A sign that reads something like "Home of the $6,000 boar. Please call after six, when we're home" is probably providing too much information—way too much information. The sign needs to attract attention, but to that end, it should neither be too detailed nor too ornate.

4. Prepare a simple one- to four-page, double-fold flyer. This is something for potential buyers to take along and study at their leisure, and it can be included in mailings or posted in public places. This is a good way to keep you presence known to the public, using a message you control.

5. Get the name, address, and phone number of every buyer and visitor to your farm. Review that list every 6 months or so, to see who might be needing a new boar, another set of feeder pigs, or pork to refill the family freezer.

A simple postcard sent to previous buyers will remind them again of you and your business. It will not be intrusive and can get them up to date on what you currently have available that might be of interest. I've sold several boars over the years to folks who told me frankly that they had nearly forgotten about me until they got a postcard that thanked them for their past purchases and asked if I could be of help again. Many people also appreciate a follow-up call to see how they are faring with their purchases.

6. Cooperate with other local farmers to draw buyers to that area. Working on such things as a simple map, shared newspaper ads, referrals to each other, and a directory, and a network soon emerges that will draw all sorts of potential buyers to the area. For some reason, farmers don't like to shop real close to home.

Some say that there is a recreational aspect to the buying process for many people in our culture. The more that can be presented to encourage them to come out, the better your selling opportunities will be.

I was away at school once when my father called to say he was selling all of the sows—it was over. In response to my anxious inquiries, he told me that that very morning he had managed to pull off the paramount act of the business. He had sold a young boar to the old gentleman that farmed across the road from us. The old gentleman was a very good neighbor, but a staunch fiscal conservative. He was right; no way could we top what he had achieved that morning.

7. There is a very fine line that farmers draw when evaluating advertising, and if they believe you have crossed it, you stand to lose their support. When you do invest in some advertising, keep it respectable: no bragging, no overstating a point or slamming the competition. Show what you have and how to find it, and they will tell you how good it is with their dollars.

8. Share your professional knowledge and passion for the work. Try to support groups like the 4-H and FFA. Donate your product to local charities. Go to the livestock shows. Wear a breed cap. Buy some of those support ads you see for area groups. Join some of those area groups (fair boards, extension councils, etc.), where you can do a bit of promotion for your style of production and your hogs. And get out there and make your skills and abilities available to others.

Phyllis and I worked with the local 4-H club for many years and enjoyed it very much. We took in a lot of hog shows and spent many long days with these folks, but the rewards were great. While we made it a point to not sell anything to the youngsters we worked with, we developed a good reputation for our knowledge and professionalism, and naturally that generated an interest in our business.

9. Bear in mind that your investment in marketing is just as important and should take as much time and resources as any other facet of production. The markets may not always be good, but in some form

they will nearly always be there. It is the producer's task to recognize and then grow those markets.

Good marketing is simply telling your own story in your own way. You reach out thoughtfully to those who will respect and reward your efforts as a swine producer. The range producer needs to tell the range production story.

Direct marketing has been proven to provide the greatest returns to the independent producer. It will place you in a one-to-one relationship with a great many buyers, but not in an adversarial way. It is actually the most natural way to sell, one you will grow at ease with as time passes.

Going into the marketing phase, the range producer needs to begin by enumerating the strengths or merits of the products that will be presented for sale. This should also increase your comfort level in this marketing process by giving you something to rely on. Learn and be reassured by these most basic truths:

1. The pork that's offered by the range producer scores highly with American consumers because it is from the family farm, is generally additive-free, it's fresh and flavorsome, and all of this is neatly embodied in that simple phrase "range pork."

2. The breeding stock you produce is durable, hardy, and ready for life in nearly any environment. You know this because you are a farmer producing for other farmers, and what you need to have to work on your farm is what they need to have on their farms, too.

3. The feeder stock you have for sale has been hardened off and will go anywhere and should work for anyone.

The selling job for any of these aspects of your business is simply to tell the truth about them, about how they were produced and about yourself as the producer. Believe in yourself, and you will naturally believe in your production and its worth to others. Straight up, if you don't tell the story of this production, it won't get told, and if you don't promote your production, you and your operation alone will fail.

Pork production may have been grappling with hard times of late, but none of this is really new. The night we bought our first Duroc gilt, we could have bought two Black Poland gilts and one boar, all blue-ribbon winners, for less than $150 for the total package. There hasn't been a lot of good news for a very long time. The support infrastructure has contracted sharply in many areas, many pork magazines have folded, producer groups are not dealing well with reform, and pork as a simple commodity has been left to a few giant corporations for processing and marketing. To keep pace with the times, the "other white meat" is now being termed the "other white protein," but it only seems to distance it that much further from the family farm.

Pork as a commodity is now a near-perfect example of vertical integration. In this industrialized process, hogs enter at one end, and pork emerges from the other. And since it's all done in one house, profit is taken at only one point—the retail sales counter. The contract producer is working for pennies and is really only left with the mortgage on the facilities and the mess. As some might say, you get a small check, the loan payments, and the slop in the lagoon.

A few marketing groups have taken up outdoor-produced, natural pork, and they have met with a fair bit of success. It is still very early in the game for these folks, and much needs to happen for them to build a lasting presence. They can produce good hogs in good numbers, they're investing in producer-held processing facilities, and they have some good pork products to sell. Getting it distributed and into retail outlets has been a significant problem, however. Distribution beyond a few local markets is a challenge that many such groups have been unable to overcome.

Any group actually attempting to go head-to-head with the remaining big packers will, I fear, be met with the Wal-Mart treatment: they will lack the resources to go many rounds in a price war and will have to spend too much money assembling shackle spaces and a delivery system. Small, producer-backed packing groups will meet up with a real beat-down, should they make even the slightest move toward meat-case space in the big-time retail sector.

At our local supermarket, a small regional chain, they maintain a display case of numerous pork sausage types and brands. Read the fine print, however, and you will quickly see how common and close to home their origins really are. Much of retail market is now a tightly closed world that will neither welcome nor even grudgingly accept anything in the way of new players.

Marketing for Hard Times

A discussion of markets and marketing would not be complete without a few thoughts on what to do when the prices are circling the drain.

Hard times are going to happen. That's true for just about anything. For a long time, too long a time, swine production was gripped by a very Darwinian mindset: the fittest would survive the price valleys, and those that fell by the wayside were lost because of their own failure to conform and compete.

That mindset lasted until a great many of those who helped spawn it began peeling off themselves into those ever deeper valleys between ever lower price rise hillocks. The game for the independents is around and in the wake of the big operators. Let them have the tapioca-bland, ball-bearing-uniform market part of the pork commodity.

The range producer must rely upon the unique and very special nature of his or her production. Range pork, pork products with a strong breed identity, classic cuts and processing: these facets of the business all have a market here, but it's not one that is recognized by the driving forces in the fast food nation. The small markets are ours if we just recognize and serve them. The factory gives us fish sticks; the fisheries folks with heart and soul give us caviar. Tons of restructured pork sold as "riblets" return little or nothing to the people back on the land.

The independents, the range producers, and others need more organization. They need a presence that goes beyond some shady checkoff-fund promotions and a few shackle spaces reserved for the "little guys." What is sorely needed is a way to fully substantiate independent producer numbers, to document how they have been treated in the marketplace, to link them together to share their efforts, and to reach out to consumers. The organization committed to telling the whole story of what pork was, is, and can be once again still has yet to be realized.

On the farm—the frontline of the low-price wars—there is a lot that each producer can do to maintain a viable presence on the land. Always be a realist and move rapidly with counter measures to avoid suffering a complete train wreck in the face of a market in free fall.

Most price plunges do not occur overnight. Often, you can hear their footfalls coming a long way off. Hot winds blow for many days before a corn crop fails, and sows have to be bred before pigs can flood a market. Watch USDA Crop and Livestock Reports, note numbers of gilts being retained, stay plugged into the marketplace at all times, chart markets, talk to people across the country who are involved with the swine trade, and then be a truly discerning user of all of this data. Never panic, never hold on to wring the last possible penny out of a market, and always endeavor to be moving ahead of the forces that shape and drive markets.

One of the truly wealthy, money-in-the-bank rich farmers in the Midwest was once asked the secret of his success. His reply was a bit cryptic, but still to the mark: "When everyone else is walking down the street, the smart money is walking up the street."

His words marked him not exactly as a contrarian, but rather someone who valued the independent course and the mindset it calls for. Markets now are going to be made one hog at a time and delivered one buyer at a time.

In a downturn, you'll want to protect your genetic base, narrow your risk, pare operation costs, and try not to despair about the course of events. Down markets, disease, and drought happen to everyone, so never draw your full identity or role in life solely from your work as a farmer and pork producer. Part of a good networking system is mutual understanding and support during hard times.

The most valuable animals on the farm are the younger members of the breeding herd. They represent the most concentrated, most advanced genetic resources and have the longest productive lives before them. Cull those animals last.

Face the inevitables in the farming life early and in a head-on manner. When the pricing storm clouds begin to form on the distant horizon, start preparing by culling the poorest producers, selling young animals as feeder stock, and putting growth plans on hold, and then swallow hard to cut as deep and early as you can. If the price dips are fairly localized, transport your animals outside the area for sales, if at all possible.

Save your choicest feedstuffs for animals in late gestation or for nursing sows and the very young, since that is where they will go the farthest and do the greatest good. Sadly, the bottom third of most herds and flocks all too often rob from the top two-thirds.

The poor performers take feed, labor, facilities, and other resources that could be better applied to the more profitable animals on the farm. Such trimming should not be viewed as a slashing of an operation, but rather a renewal of the producer's commitment to quality and to doing the best job possible. If properly timed, such sales should generate the capital needed for maintaining and upgrading the existing herd. When butcher hog prices tumble, seedstock prices soon follow.

Some of the most successful stockmen I have known would sell every head on the farm when prices were close to peaking, thus maximizing their income over and above what they had invested in the animals. They would outfit the farm again when prices were bottoming out, thus counting on the cyclic nature of livestock and meat prices. This is not an in-and-out mentality, however.

In-and-outers tend to jump on the wagon during rising prices and bail out at most downturns. They wind up trying to build numbers during a time of rising prices, but they achieve numbers and good flow only after prices have begun to slump.

A third group of farmers are simply in it for the long haul. They make their start when they have raised their initial stake, and they stay true to a long-term plan of growth and development through good times and bad. They set great store by the simple old adage which holds that the best way to have hogs to sell when prices are high is to have hogs to feed when prices are low. Over the years, income will even out between highs and lows, and the producer who is consistently in place will catch all of the highs during his or her time in production.

Every producer has tales of times when it looked like the best thing to do was to sell every head, tip the equipment, and pull all of the stock sheds up around the house. Such decisions depend upon your ability to gather and process the data you have to hand. Lately, a lot of people have backed away from hogs completely, not because of the animals themselves, but rather, because the motivation for keeping them as the primary source of income has changed. No farm or ranch should ever try to depend entirely on a single venture. Hogs are now on a long trek back to being another modest-sized venture that can be included in a fully diversified farming mix.

Some other ways of minimizing risk include cutting numbers, marketing young stock as feeders, selling butchers at lighter weights, shopping around for cost-effective inputs, delaying certain inputs, and having other income-producing pursuits around as part of a fully diversified plan of operation. And don't be afraid to be creative.

You can replace swine ration corn with other grains, such as grain sorghum. Also explore feeding options like bakery wastes. But don't go overboard in your quest to save feed dollars.

I worked in feed sales for a number of years and frequently heard a story passed around about a farmer who set about improving feed efficiency simply by cutting in half the amount of feed he gave his animals each day. When asked how that was working, he would reply that everything was going just fine. "The only problem comes 10 or 12 days into the trial," he said, "when they catch some sort of disease and start keeling over."

It's a good tale that reflects a desire we all have to hang on to those dollars they pry out of our hands down at the elevator. It is far better to repeatedly cut herd numbers than to shortchange any animals' nutritional intake. A percentage point of crude protein can be pared from time to time, but it is probably best to keep the animals moving and to get them off the farm quickly, even when corn prices are high.

To borrow a line from the consumer faction, the range producer needs to be one dedicated comparison shopper. Read those feed tags, check those prices, consider volume discounts, and wear out some pencils doing the math before making your decisions. Some feed bargains really aren't what they seem when you get into the fine print on the feed tag. Some high-dollar feeds are worth every penny if they generate corresponding growth performance.

Most equipment purchases can wait, but sometimes in down cycles, you may encounter buying opportunities that will justify tapping into that rainy-day money. Used hog equipment doesn't lend itself to being stockpiled, since much of it is like older homes, deteriorating faster when empty than when in use. Still, when you can buy something for pennies on the dollar, you have to consider the purchase.

I can get a little carried away on the subject of butcher weights, since I feel that the bastard child of vertical integration, the heavyweight butcher hog, has done us as much harm as anything else. The younger an animal, the faster and more efficiently it grows. A while back, the 220 pound butcher hog was leaving the farm at around 5 months of age. It was naturally leaner, produced handy-sized primary and secondary cuts, had good eating qualities, and didn't drown the market in seas of pork that took longer to grow and more feed to produce.

Price storms, like rain storms, will eventually pass. Some move by nearly as quickly as summer showers, and others leave you feeling you have a lot in common with old Noah and his day-after-day trials. They do occur with a fair degree of predictability.

As a range producer, you should be working with much less overhead than a confinement producer. You will have more flexibility than the producer who has contractual and debt obligations that require a constant flow of set numbers and types of animals moving through the facilities.

When I was a youngster, a severe drought cost us a corn crop and a cow herd that had been years in the building. The day after the cows went up the chute and onto the trader's truck, Dad was out the door, doing chores with the few calves and a couple of Red gilts that remained.

It was one of the hardest blows I ever saw him take, but with a large part of his dream gone and just a few animals left, he found work around the farm to keep his hands busy and his heart and mind occupied. I don't adhere to the kind of philosophy that can be condensed into T-shirt sized bites like "What doesn't kill us makes us stronger," but I believe that hard times can be endured, and personal and world history both attest that good times do come again.

Just to be safe, know your markets, know what drives them, and position yourself to address those markets directly. I have been confounded sometimes by production-ag circles that blatantly attempt to dictate what consumers will have and what they won't have. If you want a blue Ford pickup, heavy-handed attempts to sell you a red Chevrolet are only going to offend and anger you. Yet many farmers in production agriculture feel free to scoff at consumer concerns. They want to impose products that clearly aren't called for but that are more easily produced by an industrialized agricultural system.

Supply and demand and seasonality will always be at work in agricultural markets. These forces spawned a marketing system most folks were comfortable with and that was fairly rich with both variety and opportunity. Hopefully, we have not come to this realization too late to preserve and continue its best elements.

As a testament to this possibility, take a walk down the aisles of an Italian specialty food market or organic grocery store; you may be shocked at what you see. There are sausage products crowding $10 a pound, butterflied chops selling for more than prime sirloin, all of this tapping into their origins in naturalness and tradition. These sorts of things make it very apparent that the Tyson marketing model not only punishes the average producer, but denies and deprives the consumer.

To close this chapter, I'd like to note how the enduring hardiness of a single breed type has managed to thrive despite the hardships of cultural and technological changes.

The people of Britain have long prized a grand old breed of hog, the Gloucester Old Spot, and they working hard to preserve it. This is the parent breed of our Spotted hog, sharing a herd book with that breed here. The British royal family is even involved in its preservation and breeding.

This breed has occupied a very unique role in British agriculture, where it has come to be known as "the orchard hog." I recently saw an Old Spot rather prominently featured in a couple of episodes of "Masterpiece Theater" on PBS.

For over a hundred generations, these animals have pursued a course of life quite similar to that of their European wild boar ancestors. Living in part on fallen fruit, grasses and forbs that grow between the trees, household scraps, and a bit of grain, they have come to symbolize the true cottage-agriculture hog. What wonderful roasts and suckling pigs they must have produced (and may well be producing again).

They continue to exist due to the hardiness of their litters and themselves, the important role they've fulfilled there, and the value that has been placed on them by Britain's independent family farmers. A few of these animals are now being bred in this country, where they are being sought out by farmers turning anew to outdoor swine production.

Just imagine what a roomful of Madison Avenue ad men could do with the marketing and promotion possibilities of these marvelous creatures. There are so many facets to its rich history: they represent hundreds of years of history and traditional dining; they reaffirm the classic tie between pork and apples; they are humanely reared outdoors; and they evoke all sorts of positive pastoral images. There are even pictures of Prince Charles admiring them as they graze. Somebody is going to put sausage made from these animals into packaging sleeves featuring thatched British cottages or images of Pilgrims and the Mayflower and make a lot of money doing it. That somebody could be you.

Farming doesn't stop at the road gate anymore. It never really did. If range-rearing is how our grandparents did it, then let's go ahead and take a further clue from them.

Didn't they take their butter, eggs, milk, and cured meats and go to town with them? Isn't it time to do that very thing again?

BERKSHIRE SOW

CHAPTER 7

HERD HEALTH

Taking hogs back out-of-doors is essentially taking them back to their natural ways. However, the modern range producer doesn't simply open the doors of his confinement units and send them out there with a "root hog or die" attitude.

Over the years, we have found that our greatest successes have come when pursuing a more traditional path of production but still adopting new measures that show merit and promise. The good outdoor producer is far from a backward thinker, and may actually now be considered cutting-edge.

Anyone who is serious about outdoor production should focus on breaking the pattern of dependence on the practices and products that have cast such a shadow over animal agriculture of late. You cannot make a silk purse from a sow's ear, nor can you create quality pork out of a stress-laden and severely driven confinement situation.

A good illustration of this is the dearth of quality in culled sows that emerge from these units. Virtually every time that hog prices buckle, the last market segment to be sharply impacted is the sow market. At slaughter, most heavy sows are bought for the whole-hog sausage trade, since they yield large amounts of pork that have the profitable blend of fat and lean needed for that product.

These days, few are the confinement sows that last to even four parturitions. They emerge from confinement units rather like husks, weighing less than 400 pounds. In fact, they are starting to have a lot in common with the cull hens from the poultry industry. These small sows carry little in the way of cover, and they yield poorly.

Sows coming out of a more traditional environment may continue in place and producing for six, eight, or more litters in a lifetime. They will have had relatively free and active lives, receiving a diet regimen that contributed to their longevity and productivity. Both of these ultimately contribute to a substantial salvage market value once their breeding days are over.

The older, bigger sow doesn't lend herself to being packed away in gestation crates and tiny pens. She can move through some such facilities rather like that proverbial bull in a china shop. Still, she will have earned her size by being a good pig producer and raiser.

She will have emerged as a proven, dependable ongoing genetic resource within the herd. Over the years, she will have built up a strong natural immunity to the varied "bug soup" that is a part of every farm with livestock. She will then go on to impart this immunity to her offspring, beginning with a milking in its first moments of life.

There are old hands at this calling that will only keep female herd replacements from their oldest sows. These are the "gems" of the herd, if you will. Their genetics are known and proven, and they have the added bonus of that natural immunity.

As a youngster, I was once given a tour of a bona fide working farm of its day. In a far barn, I was shown a handful of nursing sows of near-epic proportions. They were huge and showed considerable age, evidenced by the fact that they were attended by fairly small litters. They were the last remnants of the sows that began that farmer's operation, and although a bit of sentiment could account for their continued presence, it was the quality of their female offspring that he prized and that had earned them their protected (and protracted) status there.

In bringing new genetic elements onto the farm, many producers will first breed those animals with their oldest lines to see what they can really do for or to the existing herd. To be sure, a good mating will produce offspring better than the sire and the dam, but this doesn't mean that sows should necessarily be dumped after every other litter. Our first gilt stayed in the herd nearly 5 years, and we were about 10 years in before her last daughter left the farm.

As you can see, swine health care is therefore not all injectables and elixirs. Health care can follow a more natural scheme at all phases of production, without impacting performance or output.

Simply by selectively breeding for wider and deeper chest cavities, greater body capacity, and well-formed reproductive tracts, you put yourself giant steps ahead of many potential health challenges. If you can follow the tenet "breed 'em tough and raise 'em thoughtful," you are more than halfway home when it comes to the question of swine health care.

The Issue of Natural Production

Raise the "natural" question in some hog circles today and you will open up one mighty big can of worms. And the two camps in which you will create the greatest consternation are veteran confinement producers and those in the organics side of the business.

There are a lot of second- and third-generation confinement producers out there now, and although they may be hurting at the moment, they are still fighting to hang on to something that is all they have ever known. It reminds me of that Zen teaching about the man caught midstream who is being pounded to death by the river's current and yet he is too afraid to turn loose and swim the short distance to the safety of the nearby stream bank.

A few years back, I saw pictures of an operation that was actually double-decking gestating sows in crates. This setup had producers reduced to taking down and putting back "swine production units" on an as-needed basis. It was industrialized farming at its finest, but even a dyed-in-the-wool anti-PETA fellow like me had to wince at those images.

The organic camp has some good ideas, but they have not been given the keys to the kingdom. In a pure organic system, ill and injured animals are going to have to be removed, because you will lack sufficient measures to counter every challenge to herd health.

Read the Herriot books to see what a real boon some of the modern products and methods were to stock raisers who had had no other options. To be sure, some drug products have been abused and rather badly, but wise and judicious use of what is available is not necessarily at odds with the goals and values of good stewardship.

The old hog has had image problems of one sort or another for as long as anyone can remember. Even those folks who raised hogs as their bread and butter felt themselves slightly inferior to their cattlemen neighbors and their charming cowboy image.

My first clear memories of the hog scene are from the late '50s and early '60s. Conditions then weren't favorable for hog production either, since overall things were on the decline then. Prices were traditionally rather low. In fact, at that time you could buy good brood sows for far less than the cost of a good steak dinner now. At the FFA sale where we bought our first Duroc gilt, the top Black Poland boar brought $65 dollars, and good gilts just a bit more than half of that. And that was the mid-'60s.

On the technology front, things had been rather at a standstill since World War II. The question of lard-type versus meat-type had been resolved, but many hogs were still carrying a lot of grease. The lard image has been a struggle for generations. It ceased being a factor when grease was no longer needed for the war effort. To steer consumers away from this image, farmers no longer use the term "lard" or even call market-ready animals "fat hogs," as was once common. The bacon and lard breeds both now produce first-rate slaughter animals with lean, high-yielding carcasses.

Still, breed gene pools were quite broad, and good hogs dotted the countryside. Thirty sows constituted a big herd back then, and diseases like Porcine Reproductive and Respiratory Syndrome (PRRS) and pseudo-rabies was unimaginable.

Some change was due, maybe even overdue, but who could envision the rocket ride up and down from '65 to '95. I was there for all of it, and it still has me scratching my head.

The drive to steer production indoors was prompted from the top, institutionalized by the ag schools, and lauded by the farm press. The reasons why the hog was taken from its vaunted role as a part of a diversified farming mix and catapulted forward as a single-venture pursuit was as much driven by issues surrounding image and ego as any other.

I believe that for some folks, the hog had come to symbolize the drudgery and the hardscrabble image that was affixed to farming in the '20s and '30s. It may have been this backwoods Okie image that motivated such champions of industrialized agriculture as Earl Butz and other equally driven men and women.

Ma and Pa Kettle may have had a hog on their front porch, along with the family hound dog, but that was Hollywood, not rural America. The hog gave balance to and paid a fair share of the bills for a fully diversified family farm. (And just between you and me, Arnold Ziffle is still one of the brightest and most erudite characters ever to appear on a TV sitcom.)

The hog factory had to be paid for, however, and as soon as it came on line, so began the long descent downhill. The drudgery of tending a few head outdoors was replaced with the drudgery and worry of tending hundreds or even thousands of animals indoors, simply to keep facilities on line and some semblance of a cash flow maintained. But there was a trade-off in the switch from single-farm-level drudgery to large-scale-drudgery, not the least of which were ever growing energy costs and dependencies, changes in swine type that broached upon actual changes in swine physiology, a variety of environmental ills, perpetual crowding stresses, and a whole raft of new diseases.

There are producers now who go about their daily chores with a shirt pocket full of syringes and a small valise with injectables for every sort of snort and wheeze. For a time, boars were given shots to make them breed, and sows were injected to speed up labor. And even now, a lot of pigs farrowed into confinement are tubed within hours of birth with a potent drug cocktail just to keep them alive.

It wasn't so long ago that hogs were actually being fed arsenic to control parasites, and to this day many types of feed are heavily laced with a number of different antibiotic combinations. The animals that once got by happily drinking from cow tracks and foraging for grubs and acorns have been reduced to what is now little more than a porcine version of the delicate African violet.

Huge numbers of animals tightly packed together and generally living always above or immediately adjacent to their own wastes creates a natural milieu for pathogenic organisms. Any health problem that gains even a toehold will remain until the facility is totally depopulated, burned clean with strong chemicals, and left to stand empty for a period of time.

There are some confinement units in which newborn and nursing pigs are constantly present, thus maintaining a perpetual point of vulnerability in the system. This is because there are a number of health problems for which the farrowing house is a gateway into the herd. Some producers have attempted to circumvent the threat by breaking up their confinement operations and spreading their systems across a number of farms some distance apart. They will farrow on one farm, early wean and move the pigs to another farm for nursery care, and from there move the pigs to other farms for finishing. This explains those cute TV news stories of little pigs running through rush-hour traffic whenever one of those transport trucks turns over. We don't see them a couple of days later, when they're dying from stress and exposure.

But try as they might to defeat disease by spreading their operations around, these producers are still trying to contain an ever surging sea of animals. The stress load remains constant, and the numbers involved at this level still ensure that pathogenic organisms have access to multiple hosts and easy transfer in between, that their life cycles can continue uninterrupted, and that there will be a continual build-up of waste loads to support them. Of course nature will not tolerate these sorts of numbers in wild populations; there disease is the natural check to an inflated and unbalanced population.

Thus, the second line of health production then is modest numbers. A healthful environment is never crowded or overwrought.

Maximum production does not equate to optimum production. In fact, on farms where livestock is produced, these two ideas generally emerge as being diametrically opposed to each other. When numbers are pushed in an effort to offset overhead and other costs, performance almost invariably suffers. Manpower, machines, animals, and lagoons can readily become overtaxed. Pushing up numbers is no way to counter falling prices, especially since the surge in numbers tends to depress prices even further.

Volume, it seems, is seldom if ever the solution to any of the problems faced by the independent farmer. Excess numbers of animals, especially those that are tightly contained, seem to draw disease problems like a cow patty draws flies.

Even back when I was known to more people as "the kid with the pretty good Red hogs" than by my real name, I could see the folly of high numbers. I began to get the picture after sitting at enough hog sales where I watched as the good pigs from just two or three big litters would sell for a third or more of the price of the whole bunch. With modest numbers, things get done right, they get done on time, and the producer doesn't have to stint on needed inputs. You won't find this in any textbook, but more hogs are lost to "too many hogs disease" than for any other reason I know of.

In the old days, such losses were explained away with the catch-all excuse, "_____ ate 'em" (just fill in the blank with your favorite local predator: wolves, panthers, bears, bobcats, or thievin' neighbors). Big litters at weaning, and overall good performance hinge on the quality and level of care they receive, which decreases as numbers increase.

In the face of disease and general health problems, the producer with modest numbers is generally able to ferret out and correct the problems much more quickly than the overburdened farmer. Today's independent farmer cannot afford to become lost in a sea of look-alike porcines. The sooner you get on a health problem, the easier and less costly it is to treat, and the quicker the animals will return to good health.

Producers working in confinement operations or those who have larger numbers accept a degree of animal loss as simply another cost of doing business. A lot of small pigs have been dispatched on the farrow-house floor over the years because producers believed that they didn't have the time to get them all raised. Of course, how they were able to reconcile this act with the one great argument made for confinement, "you have to have numbers to be a player," was another one of those things never made clear to us "common folk."

The Organic Question

The buzzword of the moment in production agriculture is indeed "organic." But as with many other types of farming, it is actually a long-established production practice. Up until at least the 1930s, most swine producers were organic producers, or nearly so.

Organic production now comes with a fairly rigid set of guidelines and requires the producer to operate according to very stiff codes of regulations and submit to on-farm inspections. Pastures will have to be returned to a state of organic certifiability, and this may take several years.

I and a great many other farmers harbor some doubts about just how easy it will be or if it is even possible now to return pasture and cropland in the Midwest and other areas to a truly organic state. The problems posed by both pesticide drift and chemical runoff just may be too great to overcome in these areas. Most farms will take 3 or more years to get online to produce organic feed grains, and all that work can be lost in a day if a neighbor picks a windy morning to haul out the spray rig.

At the moment, organic grains are available only in rather limited amounts and often must be shipped over great distances. They're expensive to buy and costly to get on the farm, and thus they add substantially to production costs. A lot of grinding and mixing equipment relied upon by independent producers also may not pass muster. There are simply too many nooks and crannies in a mill where potential contaminants can lie up and then cause later problems.

With current organic regs, there is very little recourse for dealing with sick or injured animals, other than immediately removing them from the farm. Once an animal has received nearly any kind of medication, it must be removed before it has the chance to defecate or urinate and thus contaminate the farm. Parasite control now also falls into a very gray area under the organic program.

All of this being said, however, I still believe that the hog is perhaps the best candidate among the large animals for organic production. With hogs, you must begin with animals of good type and vigor, keep the numbers low, and work very hard at keeping out any health threats to the herd. The organic pork producer will almost certainly be a seasonal producer and will be using pasture rotation to control parasites anyway.

Rations will have to be entirely plant-based, and pigs will probably have to stay on the sows for a full 6 weeks. When I was small, we had a couple of older neighbors that left the pigs with the sows for as much as 12 weeks and let the sows do their own weaning. These folks were content with one litter a year and would market the females in the fall, after they had regained flesh. This might be a practice some organic producers may wish to consider anew.

Some parasite checks can be achieved with the use of products such as diatomaceous earth, but they do lack the effectiveness and ease of use of some of the newer products. The producer who has a small land base will be especially challenged to keep internal parasites in check.

Taking animals on the range is the logical place for the organic pork producer, since the reduction in stress alone will help ensure a defense against health-related problems. The animal bred for the body capacity and hardiness required for range life is many steps ahead of the pack when it comes to maintaining good health with minimal artificial crutches.

A Middle Way

By degrees, a great many producers are beginning to see the light and are pursuing a more simplified and natural course of swine production. They are mixing together some of the many options that are available to them to create a system that works best in their regions and on their own farms.

Most begin by backing away from the use of antibiotics in feedstuffs and by selectively breeding for those traits that contribute to health, vigor, and longevity. Beyond this, they have their animals on a regular course of parasite prevention, they feed quality rations, closely monitor animal health and conditioning, and react quickly when animals manifest any problems.

A bottle of antibiotic isn't an inherently evil thing. It can be overused and thus abused, but where such products are used infrequently and to address a specific need, they perform quickly and generally with a minimal number of doses. I've certainly discarded more antibiotic product for reaching its expiration date than we have actually administered over the last few years. I try to use it early when a quick and timely response will bring maximum results. A vet once shared this bit of philosophy with me: "Hit it early and hit it heavy."

We keep a bottle of long-acting antibiotic on hand, but buy it in the smallest vial possible. No one likes to dose sows less than I do, but when I have to, I mean for it to work.

The modern, conscientious consumer wants a meat animal produced with minimal additives and treatments, although they are generally understanding that treatments are sometimes a necessary part of humane health care practices. Nonetheless, I've seen barn refrigerators that looked like they belonged in big city hospitals. There were drug bottles by the dozens and injection timetables only slightly less complicated than the flight schedules in and out of New York City. It is only a matter of time before ad lib antibiotic use in livestock will be ended by legal mandate.

In a more natural setting, the pig in its first few weeks of life should receive all of the immunity it needs naturally from its mother's colostrum, in the first hours of her milking. From that simple, truly natural starting point, much can be accomplished healthwise just by keeping the animals comfortable, providing them with dry and draft-free sleeping quarters, keeping them well fed and hydrated, and not having them live in any sort of stressful state. (Pretty much the way my wife has been taking care of me for years.)

Some Necessary Skills

There are a number of health care skills that the producer should master, but these can be learned over time and are best acquired in a hands-on manner. Some were taught to me in my high school vo-ag classes, others were learned while helping around the farm and with neighbors, and still others picked up from a few simple lessons from our local veterinarian. With my family's first large set of shoats, Dad summoned the vet and paid him extra for time spent teaching me some of the basic skills, such as giving intramuscular injections and performing castrations.

The typical farrowing and baby pig care skills had been taught me in vo-ag. But working shoulder-to-shoulder with the local vet gave me the confidence I needed for undertaking those healthcare chores for shoats and adult animals.

A certain ease and mastery of these skills comes from working with someone more experienced and willing to take the time to share them. These are important skills that are honed with practice and can be made simpler with an investment in quality implements and restraints.

Given time, the producer should master the skills needed for castration, tending newborn pigs, giving most injections, applying parasite control products, and taking body temperatures. Most can learn to pull some pigs and treat simple injuries, and even assist with farrowings in many instances.

Each producer will have to determine what skills to pursue, and not all successful producers necessarily have the same skill sets. Do only what you feel comfortable doing, and even then, only when the animals are safely constrained.

It is no sin to call for reinforcements. In fact, the faster a vet is called to the scene, the more effective he or she may be. I have had a vet walk me through a troublesome pig delivery over the phone, but it was far from the finest hour for either of us. If in doubt, ask for help. The earlier such help is summoned, the less costly it will be.

Whether we're talking about the work of a carpenter, an electrician, or stock raiser, the work will be smoother and the results improved using quality tools. The range producer should invest in the following tools and become well schooled in their use: a rectal thermometer, needle teeth clippers, ear notcher, castration knife, assorted syringes, needles, a hog ringer (to apply a nose ring), an assortment of rings, a carrying case or tray, and a couple of standard health references such as the *Merck Manual.* Buy quality tools that have as much capacity and versatility as you can afford.

You should then match your tools with a good assortment of health care products. Begin simply with an astringent wound spray (aerosol or pump), an oral antibiotic, an injectable antibiotic, vitamin/electrolyte products, wormers, dust or spray products for external parasites, and then add any special-need or prescription products. Some of these will require dry, dark, preferably securable storage, and some must be refrigerated.

I still keep some bird and mammal treatment products in the family refrigerator for quick access, but this is not the wisest thing to do. Phyllis hasn't served me any oral iron product over breakfast pancakes, but in a home with little ones or where a product can get lost in the back of the box and spills, serious problems may occur. Older, smaller refrigerator models can now be bought relatively cheaply and can be set up in barns or other outbuildings for fast, on-site access. Their contents can be made secure with a simple hasp and padlock.

Near the refrigerator, you can mount an old kitchen cabinet for use as dry storage for the health supplies that don't need refrigeration, in addition to your vet tools.

These should also be kept under lock and key. Secure them with shared-key locks to simplify your key ring and your life. Get in the habit of putting things back as soon as your work is completed, and thoroughly clean all implements before returning them to storage.

Swine health care products that come in dark brown bottles or heavy packaging must be considered light sensitive. Exposure to sunlight for any length of time can greatly reduce their effectiveness. Some powdered and granular products will quickly get damp if not well sealed before being put on the shelf. You don't have to pitch very many $20 and $30 packs of health care products before learning to read the instructions and store them properly. If you are using the dash of your pickup or a ledge in the barn for drug and instrument storage, you are not only wasting money, but you may be compromising your animals' health and well-being.

The pickup bed or an old 5 gallon bucket is not the wisest choice for transporting your health care supplies, either. I'll confess to an occasional treatment in the field with a bottle or two in my jacket pocket, a syringe gripped in my teeth, and gripper tongs held firmly in hand, but it ain't professional. A variety of trays for carrying health care products are now available that will fit securely on top of a gate. They are made of strong plastics and can be bought for a few dollars. They should be cleaned out after every trip afield and then readied for the next use.

I have also seen old tackle boxes and tool boxes used to carry equipment and health care supplies afield. The trays are handy for keeping track of small items, and many close quite securely to help with keeping things clean. They can be set outside a fence but still within easy arm's reach.

One thing that cannot be stressed enough is the need to purchase quality products and tools, to learn to use them correctly, and to clean and store them away immediately after use. I've seen farmers use the same pair of dime-store clippers to clip needle teeth, to dock tails, and even snip in ear notches. They were saving pennies but running all kinds of risks that could lead to troublesome infections.

Cheap tinny teeth nippers will often crush or crack the needle teeth rather than clip them off neatly at the gum line. This can add to handling time and may even set the stage for secondary infection.

For an investment of literally a very few dollars, you can assemble the instruments needed to do a great many basic health care functions. The above list is just a few of the basics, the must-haves. I know, of course, everyone has a neighbor that does everything with just a decades-old pocket knife, a few bits of number 9 wire, and some short wisps of baling twine. In a pinch, all of us are capable of a bit of pasture wizardry, but the preventive costs are small and the risks are too great to give anything but the best when caring for livestock today.

As you shop around for animal-handling tools, don't allow yourself to become daunted standing before a display of instruments down at the farm supply store. They're typically pretty easy to use, often requiring only a squeeze of the handle or the twist of a rod to securely establish a restraining grip on the animal. Be very careful to use them only on surfaces that provide safe footing for both yourself and the animal. And remember also that time and money invested here will ultimately make your animals safer, healthier, and more comfortable.

Your skills and confidence will increase with practice, and your methods can be expected to change over time as you acquire new skills. I had to learn a second castration method a few years after we began with hogs and have seen several neat spins on the always oner-

Simple Restraints

Sooner or later, you are going to have to work with shoats that are too small for a catch crate but too big to hold easily, or you may have to apply rapid treatment to a sow far from gates and chutes.

For the farmer, that solution is a gate-mounted pig holder. Several years ago, we bought a factory-made unit that was constructed of simple steel rods. It was rated for pigs up to 40 pounds. Later, with the help of a friend who could weld, I built one of our own design, with modifications and enough reinforcements that 60 pound pigs could be held safely and easily.

Most pig holders were first developed for holding pigs for castration. Thus, they could fit over a gate or rail and then hold the pig head-down by the hind legs. With these setups, the legs are threaded through some sort of bar arrangement, and the pig's weight serves to hold the animal snugly and safely in place. Our model presented the pig belly-out, for the castration method commonly called the show-barrow cut.

With pigs thus held, any number of health treatment practices can be accomplished. One person can catch and suspend the pigs and then have both hands free for castration, injections, and just about anything else that needs to be done.

The next question is what to do with the animal that's too big to lift and fit into a pig holder. Well, there is a sow holder—two actually.

The first is the slip noose. It began, I believe, as a way to handle hogs with ropes, but the technique got better and the materials much stouter over the years. A loop slid over a hog's snout and drawn tightly will afford some control of the animal. The rope was soon replaced with wire loops, and then wire loops were tautly drawn through a piece of steel pipe.

The modern hog snare is a cable loop at the end of a short metal tube. A quick pull of the handle draws the snare tight and locks it down around the hog's snout. Some farmers will enclose the whole snout, and some try to put it in the mouth and encircle just the top jaw. Stand a bit behind the animal and draw the head up and back a bit for maximum restraint. The snare should hold all but the very largest of hogs for a period long enough for most health treatments. It will take two people to use this successfully.

A second restraining device is the gripping tongs. These are a simple set of pinching jaws with 30 to 36 inch handles. When clamped over the neck right behind the ears, they will effectively lock the hog into position with a nerve clamp of sorts. A secondary clamping point is at the base of the ham on the hind leg.

I had one set with 30 inch handles that when shut could be held closed with one hand. I pushed my luck and gave a lot of sow injections all by myself this way, but also got my teeth rattled a time or two.

It is always a good idea to have help when working closely with large animals. Before closing in on them, be sure that you have at least two escape routes planned and are working in an area with good footing for both yourself and the animal. When crowded, even the gentlest of animals will first want to flee and then possibly strike back.

Gear up to get your task done quickly and thoroughly so that you don't have to repeat it again at a later time. The more animals are handled, the warier they are likely to become.

ous task of giving a sow an injection. I touch on some of this methodology briefly below.

Health Care Supplies

I have built and upgraded my health supplies kit over a number of years. Tools and methods will change over time, and as skills are gained more instruments will be needed. Buy the best that you can afford, keep them clean and well maintained, and carefully store them away after every use. Tearing through barns to find an instrument needed in an emergency is for careless amateurs, not the proficient farmer.

Invest in a good carrying tray to keep your tools both clean and close at hand in the field. A crushproof toolbox or two would be the ideal for storage back at the farmyard. There you should return your boxed supplies to a place designated strictly for this purpose. This should be adjacent to your medicine storage to alleviate much of that confusion when an emergency does occur.

A few simple procedures and the basic tools used to apply them include the following.

1. A rectal thermometer is quite valuable in making early health diagnoses. The more data like body temperature, general symptoms, and recent handling changes that you can supply the vet, the better and quicker he can address the problem. A hog's normal body temperature is in the 100 to 101 degree range.

An elevated temperature means the body's normal defense systems are rallying to counter an infection or harmful organism. A depressed temperature indicates a very severe problem, with one or more bodily systems generally having shut down. It is, for example, one of the symptoms of kidney failure.

Buy a rectal thermometer with a sturdy carrying case. An attached wire with a small alligator clip can be clipped to the haircoat to hold it safely in place while waiting to record a reading. Keep the thermometer clean and sterile, and store it securely in a carrying case. Rattling around inside a glove box is not proper storage for such an important instrument.

2. Castration can be done either through the scrotum or on the underside of the animal between the hind legs. This latter method has come to be called the "show barrow cut," as it leaves no visible signs of the procedure. I prefer the latter, because I can do it alone with a good pig holder.

With the pig suspended head down and belly out, it is a simple matter to position each testicle between the thumb and forefinger with a simple pinching motion. A single shallow surgical stroke over each testicle and it is then freed and can be removed with the severing of the support cords. With the testes removed, spray down the incisions thoroughly with an astringent spray such as tamed iodine or Blue Lotion. Occasionally, you will encounter a pig with an undescended testicle. It may drop later and cause problems. You should make a note of this in your records and regularly check the animal to see if it has dropped.

Castration blades exist in wide array, and many an old farmer has worked successfully for years using one of the blades of his pocketknife. It may even be the same blade he uses to cut off a plug of chew from time to time.

I prefer the hook-type surgical blades. You can buy a package of a half-dozen that will work with a surgical handle for $5 or so. Now available are disposable hook blades for a couple of dollars that are adequate for working up as many as three or four litters.

3. Then there are the shots. I realize that even discussing injections is an anathema to most pure organic folks, but there are injectable medicines that will save animals' lives without becoming a crutch for the producer.

Keep an assortment of nylon syringes in various sizes on hand. We try to keep 3 cc, 10 cc, and 25 cc units around. They are inexpensive and have a smooth and fast action for rapid and less-stressful injections. When giving an injection, use the shortest and heaviest-gauge needle possible, for the same reason.

I like the disposable needles, because they are inexpensive enough to discard after a few uses, should they become dull, bent, or develop a burr. Remember, the trick to successful injections is "in and out as fast as possible."

Most of the shots you will be called upon to give are intramuscular. From little pigs to large sows, the best sites for these injections are in the long muscles on either side of the neck. This is generally an easily accessible point. Moreover, there is no staining of economically important meat at this spot, and if a needle happens to break, it is fairly easy to retrieve there. The ham is a larger target, but it is the most economically valuable site on the meat animal, and nerve damage can occur there.

A good approach is to walk slowly up to a large hog with the syringe held parallel to the top-line. When the neck muscle is reached, the syringe can be quickly raised and the injection given with one swift motion.

4. The producer on range should certainly become familiar with the longtime practice of ringing hogs. Nose rings deter hogs from much of their damage-causing rooting activity, although it will not deter them from rooting in soft mud. Hogs have also been known to throw rings from time to time.

Standard rings are firmly clamped into the gristly tissue at the tip of the hog's snout. This is a largely painless procedure, and the number of rings applied will vary with the size of the animal.

I've seen a few sale-barn rogues over the years with enough metal in their noses to fabricate a full suit of armor. On young pigs, one ring in the very top edge of the nose is generally enough. With older shoats, two additional rings—about $^1/_2$ to $^3/_4$ of an inch to each side of the center ring—will be in order.

I have seen three- and five-ring arrays used on sows. Sow rings are also considerably larger than pig or shoat rings. Never ring breeding boars, since they use their noses to position females for breeding. Growing boars can be ringed, but like hogs going to show, they should have their rings removed before being relocated (clip them free with a small bolt cutter).

A better choice for large hogs on range might be what are termed "humane rings." These clip into the very end of the snout spanning the space between the nostrils. These are far longer lasting rings than the simple crimp type, but they do require a special ringing tool. A ringing crate or tongs and a helper will be needed when ringing large hogs.

In time you will require other, even more complex skills. You may even be surprised at what you can accomplish. Once, with the vet on one end of the phone and my grandmother holding the other, I pulled a litter of pigs from a sow with an injured vulva. I saved the cost of a field call, the litter, and the sow, but that night I noticed my first gray hairs.

Working with a Veterinarian

The anecdote above brings up the point of how important it is to develop a solid working relationship with a local veterinarian. Our vet was in my high school graduating class, but the relationship needn't go back that far to be a good one.

The relationship does, however, have to be rather clearly defined and ongoing, for any number of legal and business reasons. To acquire the use of some products, particularly for testing purposes, and to acquire transportation documents, the relationship between veterinarian and producer must be clearly delineated. The vet should be familiar with your farm, your feeding practices, and your style of production. Our vet knows we are an outdoor operation, and he is also aware that I am cost conscious, and makes his recommendations accordingly. More important, he knows my skill level and what I can and cannot handle on my own.

Over the years, our vet has been able to talk me through a number of health matters, keeps me up to date on the animal health scene. He has come to feel comfortable supplying me with prescription drugs and has become a sounding board for everything from seedstock selections to ration choices for our operation. He serves as one of those impartial observers all farm operations and farmers need.

Veterinarians have taken a bad rap as the costly route to health care for the small farmer. Indeed, there are cheaper sources of health supplies to be found: there are some regional salesmen who are competent to advise in a number of production areas, and there are some quite extensive health manuals being written for the layman. Just don't expect to depend too much on those sources when you have a sow in real trouble at 12:45 on a Sunday morning.

I buy from the supply houses and route salesmen just like everyone else, but in a moderation borne of some experience. Health care products from these sources are often nearing their expiration dates, they are sold only in large volumes and have shipping and handling costs that must be factored into their final price. From the local vet, you will find that the products are fresh, they are available in a great variety and can be purchased in modest amounts, they come with his or her expertise, and often are only minutes away when needed.

There is also a bit of quid pro quo to be honored here. We've all needed a vet in a hurry or at an odd hour, or both. A regular client could and should get more consideration than someone who calls upon a vet only to put out health care "fires."

Furthermore, a lot of the added costs of veterinary care are often of our own making. Should a health problem emerge, be prepared to assist the vet by giving a detailed account of how the problem has developed, what treatments were given, relevant facts concerning the herd history, and the like. Also, be there to assist the vet with animal calls and to provide anything that might be needed, such as hot water.

Before the vet arrives, have the animals that need attention moved to a safe and secure area. This area should be removed from other hogs, have tall and solid fencing, provide easy access to the animal, and provide good footing for both animals and humans. Also, try to get the vet out as soon as possible and at a reasonable hour.

You can try dragging your feet with an ill animal, believing that time will heal whatever ails it, and then call a vet out in the dark of night or out to a far pasture, but don't be surprised by the bill. Unless your vocabulary needs the addition of a few new Anglo-Saxon terms or else you just want to test your vet's degree of dedication, try to avoid calling him or her out to the farm at dusk on a snowy evening to treat and cut 50 head of sniffling 100 pound boar shoats running through 20 wooded and hilly acres.

You should know in fairly short order when a health-related problem is beyond your skills and resources. Call the vet early in the situation, when the animal still has strong reserves to draw upon. Treatment should be less costly then, and the response quicker and more assured.

Health Care Through the Years

Health care should be a part of day-to-day responsibilities, and many practices can be penciled in as a part of a regular production calendar. Through a production season, your treatment and care log will look something like the following.

1. Three weeks before putting the breeding pen together, individually evaluate all breeding animals for soundness and general health and conditioning. Flush the animals—that is, increase the feed to improve body condition—if necessary. Worm the animals and treat for external parasites. Log into your herd book dates when all health practices are done and what health care products were used.

2. Continue to monitor herd health and the progress of the pregnancies. Many farmers will separate boars from the sow group after the timeframe for two heat cycles has passed.

A. Remove females that have not bred in sync with the group. There is a hard choice here whether to cull any "open" females or drop them back to the next group in the breeding rotation. Open sows, after two chances to breed, are probably best culled. Still, if they represent valuable genetics or there is a good explanation, then it is more of a judgment call for the producer.

B. Feed gilts so that they gain at least 125 pounds during their pregnancy, and feed sows to add at least 75 pounds.

C. In the last third of the pregnancy, it may be advisable to step up both energy feeds and protein levels by as much as $^1/_4$ to maintain the desired growth curve. This is the time when fetal pig growth is greatest.

3. Seven to 14 days before farrowing, treat the sows again for internal and external parasites. Rotate such treatment products often, to prevent any possible immunity problems from developing.

4. If possible, withhold sow feed for 24 hours before and 12 hours following farrowing.

A. Add bulk such as bran to the farrowing ration if constipation and milk flow have been a problem.

B. Offer modest amounts of feed following farrowing, and gradually build back up to full feed (approximately 3% of bodyweight) by 10 to 14 days following farrowing.

C. Provide the sow with plenty of clean and fresh drinking water throughout the farrowing period. Some farmers provide warm water and perhaps water with Epsom salts for the first drink following farrowing.

5. As farrowing time approaches (between the 110th and 117th days following breeding), observe the sow's progress, without becoming too intrusive.

A. During this time, the sow will move into something of a natal state, perhaps seeming a tad agitated. For example, she may roam a bit, remove herself from any group, engage in nest-making activities, circle and sniff often at her hindquarters, and there may be a slight vaginal discharge. It may be necessary to confine her to a farrowing hut with hog panels if she is on range. Some sows have a nasty habit of heading to deep brush or to low, wet places as the time to farrow nears.

B. Signs of a problem during farrowing include a bloody, cloudy, or foul smelling vaginal discharge, nonproductive labor for an hour or more (the first pig should arrive about 20 minutes after heavy labor starts, and the rest should appear at 5 to 20 minute intervals), and constant up-and-down behavior and other obvious signs of pain and distress. With an hour of unproductive labor, it is time to intervene or summon help.

6. Within the first 12 hours of farrowing, it is time to begin hands-on pig care. Early steps include nipping the wolf teeth, treating navels with a drying astringent, docking tails if so inclined, ear notching, and giving iron injections.

A. Observe the pigs carefully to be sure that they are getting off to a good start. The sow will summon them from the pig bunk with a soft grunting, and they will nurse on the order of 16 times a day. Healthy pigs will appear full, they will sleep fully extended and spread out a bit, and will leave the sow for the pig bunk when full. On the other hand, pigs that are piled together, that shiver, have their hair on end, and are squealing and/or trying to force the sow to nurse are in distress.

B. A sow that often lays on her belly, has vaginal discharge, is off feed, has a hard and swollen udder and/or a fever is in trouble, and her pigs are probably not getting adequate nutrition. All of the placenta should also have been dispelled within 12 hours of farrowing.

Don't go to injecting antibiotics and oxytocin willy-nilly, however. Get with your veterinarian and work out a complete plan of attack for this situation, which can crop up from time to time. I also believe that you should keep this sort of thing in mind when buying animals, to selectively breed for pigging ease.

C. The heavier the pig at birth, the better its odds for survival. Pigs at 3 pounds are ideal, and those under 1½ pounds will have the hardest row to hoe. Very large pigs will often suffer a great deal of birth stress, however. We have had some 4 and 5 pounders hit the ground running, but also have had a lot of pigs that size that were stillborn.

D. A sow that is having trouble farrowing can sometimes better cope with farrowing stress by having all of the pigs withdrawn for a few hours (up to 4 or so). Remember that just about any drug product that will calm down an edgy sow at farrowing is also apt to bring her labor to a halt. Keep the pigs warm and dry and get them back to the sow as soon as possible for that all-important exposure to the colostrum.

Colostrum isn't exactly milk, but rather the nutrient vehicle through which the sow is able to pass much of her built-up natural immunity on to her young pigs. It helps considerably to safeguard the pigs during the first 6 to 10 weeks of life. It is generally produced by the sow only in the first few hours following farrowing. A pig that

Cross-Fostering

With very large or uneven litters, it may be necessary to move pigs from one litter to another. This is a practice known as cross-fostering.

Sows generally have a very agreeable nature; in fact, some cross-fostering occurs naturally with sows and litters living in group situations. We had an old Chester sow that would literally disappear under a horde of pigs from all corners of the lot every time she lay down to nurse. But we've also had sows that would brook absolutely no intruders when they nursed their litters.

Thus, the first rule of successful cross-fostering is to know your sows. The touchy ones are seldom good candidates for foster-moms.

To successfully cross-foster pigs, the following steps should prove helpful.

1. Before cross-fostering, be sure that all pigs are clearly marked and can be traced back to their litters of origin. This ensures that their breeding background remains verifiable.

2. Cross-foster only between litters that are close in age and have pigs of similar size. The sows should be at the same stage of milk flow. First-litter gilts are probably not the best choice for foster mothers.

3. Contrary to what many would think, the largest and strongest of the pigs are the ones that should be taken from the litter. This then reduces the stress and competition for smaller and perhaps weaker pigs. Some farmers have been known to remove the smallest pigs from a group, placed them with a sow that's well known for milking and mothering, and let them nurse for perhaps as long as an extra week.

4. Use scented talc or other scent-masking product to give the new arrivals and the other pigs in the litter the same odor. This can go a long way towards helping them to be accepted by the sow and the rest of the litter.

5. Try to introduce the pigs when the sow is nursing. Handle the pigs with a light touch, and keep them from squealing or producing any other potentially stressful behavior. Carefully supervise the introduction, and be prepared to remove the pigs quickly if the sow isn't receptive.

6. Don't overwhelm the sow. Take out and replace pigs until the litters have roughly the same number as when you started. If you are transferring pigs into a small litter, do it very soon after farrowing, to maintain good udder development and utilization.

7. Recombined litters need to be monitored carefully for several days to be sure that no pigs get neglected or fall further behind.

doesn't receive colostrum is not apt to survive. Now a number of milk replacements and even some colostrum substitutes are available commercially. Pigs can be given the colostrum from other sows in the herd, if necessary.

7. Monitor the little pigs for failing health, diarrhea, and abrasion injuries such as raw knees and navel sores. Some of those injuries can grow into quite serious infections.

At some time between 10 and 21 days of age, it may be necessary to give the pigs a second iron injection. It is the lack of iron that keeps milk from being the perfect food for the young pig. Iron supplementation can also be accomplished by scooping up a shovelful or two of clean sod and dropping it into the pigs' area every couple of days. I even like to give some clean sod to sows that have been penned up for a time.

8. The time for castration can vary anywhere from day one of life to up to about 65 pounds or so. The younger the pig, the less the stress load on the pig and the producer. It is a bit more of a precision task with the very small pig, however.

9. Pig starter and early grower feeds all seem to come with one sort of medical additive or another now. Some pack very little punch, while others require a vet's prescription and must be special ordered. Some of the latter can, if purchased by the ton, cost as much as a used car. You can shop around for feeds with the least amount of medical additives possible, and you may find that a growing number of rations are being offered additive-free.

10. As weaning approaches, keep pig stress to a minimum as they move through this potentially trying time. Keep them dry, well bedded, and as comfortable as possible.

11. At weaning, some producers will also worm and castrate the pigs, believing that it is best to get everything over and done with at once. Some will give the newly weaned pig a week or so to rest and recover and then stagger such health care measures. I've done both and believe it is best to be guided by the general condition of the pigs and the weather. Don't work them up ahead of a winter storm, a cold snap, or a stretch of rain.

12. Shortly after weaning, worm and treat the sows again for external parasites. Lice can be a problem in even the coldest weather, since the animals lie so closely together. In cold weather, use a dust-type louse control product and be sure to supply it liberally to the bedding.

13. Growing hogs will probably need a final worming at around 125 to 150 pounds. If they slow down a bit in growth close to market weight, they may need another worming then, however.

14. Above about 175 pounds, be very mindful of the withdrawal times on any health care products that are used.

15. Some regions of the United States will present special parasite problems not seen elsewhere, and they are best dealt with by consulting experts familiar with the area.

16. A wide array of disease problems can confront the swine producer, and it seems that now there's an injection out there for every cough and sniffle. Still, I believe it is best to wait until you actually have an identifiable problem and then treat it head-on rather than use a shotgun approach with preventative medications. An animal with a system full of preventative chemicals may not respond as desired when treated for an actual problem.

Rely on your veterinarian and local extension agent to help you formulate a health care plan that is tailored to the local conditions. They know best what you are apt to encounter on the local scene and given the particular health history of your farm. These health treatments must be carefully thought out, since there are some practices that once begun can literally never be stopped.

Be forewarned that every farm with livestock has its own unique stew of "bugs" and vermin. Given time, the hogs bred there will build up a natural immunity, but this will of course be lacking in hogs brought in from the outside. This is the reason that over the years, many producers have established what are termed "closed herds."

Simply put, these are herds that, after the initial purchase of foundation animals, receive little if any input from outside genetics. A new animal, generally a breeding male, will enter the herd only after extended isolation and observation. A closed herd can become genetically static or even regress if not carefully managed. There is considerable discussion about the number of lines and genetic diversity to use in founding a closed herd. Close inbreeding will bring out all of the genetic good in a strain, and likewise all of the bad. To do this successfully, you must have a good understanding of your animals, your suppliers, and your genetic histories.

Health Security

One term being heard more and more in livestock circles of late is "biosecurity." It is, in effect, a concerted effort to keep out potential health risks.

In some systems, delivery trucks do not have access to livestock facilities, and visitor access is strictly controlled. No one is allowed in without changing footwear or walking through a disinfecting foot bath. Some farms actually require a full change of clothes and that visitors shower after arriving and before leaving.

I wouldn't relish the idea of trying to raise hogs inside in an armored fortress, nor I believe would most other folks. There is a lot that can be done to improve health security without escalating to such strict tactics. Steps that you might wish to consider would include:

1. Don't allow livestock species to freely commingle. This would discourage some harmful organisms that have to move through multiple species to complete their life cycles.

2. Make sure the farm pets stay close to home. Range producers may rely heavily on guardian and herding dogs, but these are stay-at-home pups. If you have a barn cat or two, show that you truly care for them by keeping them well fed, giving them good health care, and by keeping them at home. I would not care to live without a farmyard dog, but I need that animal to be obedient and a non-roamer.

3. Set up a good control program for rats and mice, including feed storage systems that keep feedstuffs from becoming contaminated with feces.

4. If farrowing in a central facility, screen out sparrows, starlings, and other birds to block the spread of health problems like TGE (Transmissible Gastroenteritis Virus).

5. Keep the hogs out of their drinking water. Hogs suffer from an "everybody into the pool" mentality, even if the pool is simply a 3 gallon pan.

6. Don't allow feed delivery trucks into lots and pastures. I've seen years when you could follow the course of a disease through the countryside simply by tracing feed truck delivery routes.

7. Post signs to deter unaccompanied visitors from entering hog pens and buildings. I sold feed for a number of years, and during farrowing seasons, we made it company policy not even to step down out of the truck unless invited.

8. Do not wear work clothes and shoes off of the farm. (A hidden bonus here is that it gives hog producers a bit dressier image when out in public.)

9. Keep all animals a minimum of 12 to 15 feet away from perimeter fences. This is because the airborne spread of disease organisms generally does not exceed 8 feet.

Producer Health Sense

When it comes to the subject of hog health, I am reminded of Dad's oft-made observation, "Don't buy somebody else's problems."

I was raised as a bit of a sale-barn rat and can recall sitting ringside all but shuddering at some of the purchases some folks made. Granted, there are "junk" buyers who make a pretty fair living reclaiming livestock wrecks, but they have that rare talent needed to pick out the ones that are still salvageable.

I have seen literally thousands of head of feeder pigs sold on the advice of the auctioneer saying, "Boys, all they really need is a little wormer, a little straw, and a little more feed." Some of them did, and some of them needed that and a whole lot more.

The key to good health care is to quickly and decisively detect emerging problems. No one can commit every symptom to memory, and many health problems can and do mimic each other. But keeping a watchful eye out for problems is the first step.

Early signs of a problem can be quite subtle and may be evidenced by changes in behavior or simple displays of discomfort. Animals that are slow off the beds or that distance themselves from the group should be observed more closely. Diminished appetite is another good sign that something may be going wrong. If it doesn't eat, it must be sick, or else it ain't a hog. These are all signs that it is time to get out the rectal thermometer and maybe start leafing through the health manual.

Be attentive early and late, when they're lying down and getting up, to really see them and get a feel for their condition. Focus all of your senses on them and note any behavior that seems out of character. Monitor wastes, noting color and texture of the stools or too frequent urination. Watch the breeding pen for blood or staining on the sows' backs (signs of high riding, failed services, or penile injuries), blood in the ejaculate, which will prevent pregnancy from occurring, and fluids pooling in the penile sheath and forming pockets where problems can form.

Walk among the hogs at least a couple of times a day. Move slowly and pause often. With large numbers on range, pick out a few distinctly marked individuals as touchstones for monitoring numbers. If your 10 regulars are there and are in good shape, the other 50 probably are too.

Move the sub-par animals out of the group as quickly as possible, but in a very slow and gentle manner. Transfer them to a secure, well-bedded house and pen, and begin the examination with the rectal thermometer to verify your suspicions. Also look for and log any other changes you can detect.

Offer choice morsels of feed to perhaps spark the appetite, and provide drinking water fortified with a good vitamin-electrolyte product to help the ailing animal to maintain good conditioning. Before drawing up and injecting an antibiotic "cocktail," consult your vet about some of the health problems that are going around and basic treatment he or she might suggest. Your early efforts can thwart or even prevent later, more effective treatment if they were misdiagnosed or poorly applied.

A healthy hog stands out like a new penny in a pail of spring water. They do indeed shine, they are alert and are curious about the world around them. They may bolt when you first approach the pen, but they will soon be back to check out your activities. I've had pigs pull on my pants cuffs like young puppies.

You can breed in traits that are important to good range hog health, like lung and body capacity. You generally foster good health with fresh and properly handled feedstuffs and clean drinking water. Everything from good fences to working the animals with a gentle touch will also add to their health and well-being.

Biologically controlled environments and drug cocktails can and do become crutches for some otherwise poor performing hogs. The outdoor producer is in essence working with a somewhat controlled form of natural selection. Over the years, nearly every breed has been impacted by a boar or two that would not have lived to breed on in an earlier time or a more natural environment.

The range producer must cull for hardiness and vigor, and some "pretty" hogs may have to be let go in the process. Actually, the ones that look really good may do so because of the fact that they aren't really working, but are simply maintaining good conditioning. An eight-litter sow with a 10-plus pig-weaning average behind her may not be pretty, but she is beautiful on paper—the bank book, that is.

A healthy swine herd hums along like a tight motor hitting regularly on all cylinders. The good producer is thus in tune with the very hum or rhythm of his or her herd of animals. Sometimes health problems are detected when the producer is just sensing that something isn't right. This type of relationship generally comes with modest numbers targeted for optimum quality rather than all-out production. High-number houses aren't what make a producer successful. Quite the contrary. That sort of system all too often enables a mediocre producer to do a mediocre job with just that many more hogs.

With 25 sows in our herd at its largest, Dad would sometimes pass me doing morning chores and say something like, "The old granny sow's big gilt didn't look just right this morning." In technical-speak, he would be saying that the litter-one gilt from the foundation sow and the Black Jack boar was a bit listless and stood back when the others in her pen grouping ate.

I preferred his way of seeing them, and I guarantee that he caught the problem at least 12 to 24 hours earlier than most of those who use the latter terminology. He had the stockman's eye and the herdsman's heart. And those are skills that are earned, not taught.

The Dollars & Sense of It

We have come to a point in time when the consuming public has some very legitimate concerns about animal health care issues. Mad Cow jokes on late-night TV and industry exposés on the news programs have us all a bit more concerned during our forays down the supermarket aisles.

Like nearly every other farmer that I know, I have stuck to this calling through good times and bad, as much for the sense of freedom it gives as the money it has put in my pocket. So of course no one wants to farm while people are looking over your shoulder, but we are producers, we want their money, and thus we should take their concerns into consideration.

Repeated polling and surveys have shown that American consumers hold not just a positive, but even a very favorable image of the traditional American family farmer. They acknowledge that farming folks have had problems of late and so are willing to invest in them using some of their purchasing power. Pork from outdoor-reared animals has clearly been produced in keeping with the expressed desires of most modern consumers.

Many consumer concerns are quite legitimate, and the farm press and educational and commodity-promoting infrastructure has failed to keep alive lines of communication between farmer and consumer. The outdoor producers have the product that consumers clearly want most, but they have had problems getting the product, along with their message, through to the end consumer. The range pork producer and the informed consumer are on the same page far more often than not.

Neither one of these groups wants to see the animals in question suffer or be diminished in any other way. Part of the task of becoming a successful range producer is to admit that the informed consumer is needed. Their wants and their beliefs as to how meat animals should be produced create the economic rewards for the range-based operation.

It is in the area of health care that they have perhaps expressed their greatest concern. Thirty-five-plus years into this, and I have to say that, yes, I have seen drug abuse on the farm. I've put a few hard-earned dollars in a couple of juiced boars that fell apart within a very few months. In hindsight, we were the luckiest with those that broke down before they were able to get many females settled.

Nearly everyone is in agreement on the prudent use of health care products that will ease suffering and speed healing. Still, overuse has brought us to the point that many now view the meat counter as an antibiotic-saturated environment. The heavy-handed use of antibiotics by the meat industry appears to have brought us to a point where antibiotics are badly compromised as a key component to our own health care.

When an animal is selectively bred for life on the pasture or drylot, a positive residual effect is that it is also being selectively bred to be a more vigorous, hardy, and healthy-looking animal. When I was a youngster, my family ran a neighborhood grocery in St. Louis County. They often did their shopping for the business and for our own table at a large central market where a variety of meat products were sold from stalls run by people marketing largely in the traditional butcher-shop style. High up in the building were large pictures of farm animals in very natural barnyard and pasture settings.

These huge near-dioramas of farm animals in open-air settings were very effective selling tools. They created positive feelings about the meat items being offered and helped to reinforce a farmer-to-consumer connection.

(This is a bit of an aside, but one of those wonderful pictures got Dad off of a bit of a hook. One picture was of a very large sow appearing to look around the corner of a bright red barn to see a litter of multicolored pigs nestled in a bright straw nest. I was at just the age that I was asking all of those embarrassing questions about how things happened and where babies came from. That great old picture was used to illustrate Dad's answer to that question about babies. He told me that sows and cows found their babies rather like my sister and I found Easter eggs.)

The importance of the range product and the healthful and wholesome consumer image it evokes was recently validated by an advertisement in the weekend edition of the *Wall Street Journal.* The ad was for bacon selling for a bit over a dollar an ounce. What could make a basic product like bacon be worth over $16 a pound to consumers? The answer will particularly surprise those who view range production as nothing but a step backward: it was the taste. The whole selling point of bacon at better than $1 per ounce is that it "tastes like what bacon used to be."

HEREFORDSHIRE

CHAPTER 8

THE BRAVE NEW WORLD OF RANGE PRODUCTION

As an outdoor swine producer, you will soon find that there aren't any "Range Raiser" magazines that you can have delivered to your mailbox, no Extension agent schooled in this type of business and eager to assist you, nor any chain of buying stations open 24/7 just to take in your farm's production.

Yours is the business of not exactly reinventing the wheel, but rather of redefining hog production for a new day and a new role in agriculture. Commodity pork production could jump to some Third World country tomorrow, and few of us would really be surprised. They could feed them out near the port cities of South America on cheap grain from the Matto Grosso (where any kind of chemical use goes), butcher them on ships heading north to the United States, dump the tankage to the sharks, and feed America pork the Wal-Mart way (*i.e.*, ultra-cheap).

This won't be your grandmother's Easter ham, and there, friends, is where the opening for the independent range producer now lies. Very little money will ever be made farming for the global village, but the folks in the big houses on the hills above the global village, now there's your market.

There is an old adage that says something to the effect that we spend our whole lives working to get back home. I think that American pork production has now made a turn and is on the way back to its roots and better days. The hog raised in simple circumstances will fit in just about anywhere, whereas the hog that serves as the sole force for a massive farming venture is overkill; it's asking too much. No other industry would limit its output the way those in production agriculture have chosen to narrow their focus.

It is only fairly recently that the livestock specialists have emerged and the hog has become far removed from its classic role on the family farm. Hogs in modest numbers were a common sight when I was younger, when 50 sows were considered a big herd. Back then, they were a valuable and well-managed part of farms that were designed to balance space, labor, facilities, and income needs.

The range producer is making that work with hogs again. The "dirt hog" has come back home.

I first heard the term "dirt hog" several years ago to describe young boars of service age. Young males grown out on dirt have few if any problems making adjustments to life in the breeding pen. This isn't so with confinement animals. It was once a common practice to move young males grown out on concrete to dirt lots for at least 30 days prior to offering them for sale.

This was to allow them to become sound again on a more natural and forgiving surface, to let some of the consequences of life in confinement (floor rubs, abrasions, pad damage, etc.) wear off, and to weed out those animals that would not become sound. A good many hogs on confinement would otherwise break up structurally (some still did). Buyers quickly learned that young boars coming out of confinement might need as much as 30 to 45 days to achieve the proper condition for breeding. Those were long weeks when the boars had to be housed, fed, and tended, without returning anything to the operation.

To the prospective boar buyer, the "dirt hog" was indeed a good thing. They still are, and now they're even harder to come by. The fit and ready boar or gilt simply represents good economic sense. In fact, it was the failure of young purebred animals to be up and ready to perform that turned so many producers away from the use of purebred breeding stock in the first place.

The term "dirt hog" is now being used more and more in the marketing of meat animals. It won't fly yet in the big city, but out in the country, where there is a strong, growing market for meat animals, "dirt" now signifies gourmet fare. The dirt hog is free of the taint and stench of confinement rearing and is less likely to have been fed an additive-laced diet. It produces pork the way farmers know it should be.

Into a Yugo world, the range producers bring Cadillac craftsmanship and values, with an emphasis on quality over quantity. A real concern for and responsiveness to consumer values and issues will likewise motivate them. These producers believe in advancement by fostering cooperation between themselves, rather than always feeling that the natural relationship is competition.

I can close my eyes even now and hear my grandmother saying, "What will be will be." Change is inevitable, of course, but what we have been told to accept as progress is not. The swine industry has lost tens of thousands of souls, believing that somehow that was the way it was supposed to be. We seem to be willing to let the people who brought us the Exxon-Valdez disaster, the Love Canal, and Three Mile Island dictate to us now how and what we should eat. And we may be only a couple of generations away from not knowing or even caring what goes into "pork nuggets" and "pig strips."

Fortunately, it would seem that the angels are on the side of the family farmers and range producers. While the latest developments in the industry may not lead to a sea change in agriculture, there is now at least something in the wind that promises better days for both producers and consumers.

In the November 2002 elections, many things came to pass, but deep in the old South, one measure that was rather quietly enacted has sent ripples wherever corn is grown and hogs are raised. The good people of Florida, quietly and with little fanfare, voted to eliminate the keeping of bred sows in gestation crates.

It directly affects but a small handful of producers—some think perhaps no more than 10 in the whole of the Sunshine State. In light of this measure, two of the farmers affected have opted to quit pork production entirely. Couple this with a recent trial and jail sentence for a confinement-unit operator here in Missouri, and many folks are questioning anew just how much of a future confinement production as we now know it will have in the United States.

The heavy hitters in the pork commodity business have tried a lot of end-runs around environmental and public health measures enacted across the country. They have tried "on paper" transfers of animals to contractors, trying to have them designated as independent producers to get around numbers limits. In most instances, this is ultimately determined to be a violation of both the spirit and the letter of the law.

A corporate move that gets a little trickier is the manner in which they play the CAFO numbers for animals in confinement as set down by the various departments of Natural Resources. People who think that 10 day weaning was just one more part of forward-thinking swine management should know that pigs 10 days old and under do not impact CAFO numbers in many state laws. You can pack a lot of sows on a farm where the "early weaned" pigs are all removed from the premises before that crucial 11th day of age.

Early weaning may thus be little more than producer sleight of hand to wring maximum numbers from minimal space even in the face of environmental measures set up to do otherwise. This is one of the dark sides of the modern numbers game.

The confinement production of commodity pork is starting to hit the wall in many areas, and all at the same time: litter size at weaning is starting to plateau at a bit over an eight-pig average; back-fat averages are creeping back up toward 1 inch to get them back to being real hogs with real flavor; performance testing is being downplayed, since it has started revealing that time on feed is edging upward; housing and energy costs now rival feed costs on many swine farms; and "the other white meat," after millions of dollars in checkoff funding (in which producers must pay into a national marketing program) and serving as the punch line of many jokes, is still largely identified as a breakfast item.

Fast food chains like McDonalds are being challenged as never before by consumers who want to see more environmentally and humanely focused choices in their meat purchases. And if you don't think this matters very much to the farmer, just realize that the simple addition of a slice of pork barely bigger than a silver dollar to just one of their breakfast sandwiches can add several cents per pound to hog prices on the hoof. Producers can profit or perish depending on the addition of only a few bits of diced pork to a breakfast burrito.

The "other white meat" now "other white protein" note raises another interesting challenge to the pork as a mere commodity crowd. Recent court action against the pork checkoff fund has greatly reduced the funds for commodity promotion and de facto support of a confinement-industry defense fund, of sorts. Still, it was an important victory for independent producers struggling to flex any remaining bit of muscle. Many farmers honestly felt that one of the uses of checkoff funding was to force-feed them the idea that they must accept the inevitable loss of their role as hog farmers.

Some would say that these are all signs of a maturing pork industry. Such things as static demand, technology that is starting to run out of steam, and growing image problems stemming from environmental, labor, and health questions will not be overcome, even with a total retreat by the pork industry to the Matto Grosso.

The hog corporations know that for every fading and peeling bumper sticker out there touting "the other white meat," there are at least three more that read "Buy American!" Country-of-origin labeling and a growing concern over humane issues in the fast food industry have to be factored into any decision for a major swine pullout from the United States. Should Mad Cow or anything even closely resembling it pop up in Central or South America, Iowa will fill up again with hogs (without even having to hang out an "all is forgiven" sign).

Recent data published by Brianne Grenard, an animal science researcher at the University of Missouri, gives the bona fides to some long-held beliefs about meat hogs farrowed and finished out-of-doors. Using a seedstock-company line of hogs built on Yorkshire, Landrace, and Duroc breeding, 48 weaned pigs were divided evenly, with one group placed in a confinement unit on slatted floors and the other placed on alfalfa pasture.

Some of the pigs had been farrowed out-of-doors in huts, and others in a controlled environment in farrowing crates. The test ran from late February to mid July. All pigs were fed a milo (grain sorghum) based diet, and all were slaughtered on the same day.

As might be expected, feed efficiency was better for the pigs fed in an indoor environment. This is the one enduring plus for confinement-reared hogs. It comes about simply because their exercise is greatly reduced and there is no exposure to temperature extremes or fluctuations. All they have to do is eat, drink, and defecate, and always in the space of a very few feet. Meanwhile, that little disk in the electric meter is going round and round, and those propane trucks are coming and going.

Pigs farrowed out-of-doors were found to grow better in both environments, however. The genetics and the rations were the same, so the advantage clearly had to lie elsewhere. They were born into a challenging and stimulating environment. And to be totally candid, they had an element of natural selection for hardiness and vigor at work in their lives, which is not experienced by pigs born in confinement. Within the space of a very few generations, this will become an ingrained trait, and one no longer showing an attrition rate.

The loin-eyes were larger in the pigs finished outdoors, and carcass weights at harvest were heavier as well. Simply put, muscles that are used develop to their best potential. They are naturally toned, if you will.

The outdoor-farrowed and reared pigs produced pork of better color and greater tenderness. They are from an environment that is as close to stress-free as possible and that still keeps them out of the road. The dollop of Duroc breeding should also have contributed some to those qualities.

The pigs farrowed out-of-doors had a greater average daily gain. This bit of data flew in the face of a lot of "conventional confinement wisdom" (an oxymoron, if I ever heard one). Supposedly, when they're outside, most of their nutrient intake should go to simply keeping them alive.

Modern range production does not mean raising hogs on bare hilltops and in the presence of marauding predators. Quality feedstuffs, clean and dry bedding, a rich environment, and, wonder of wonders, growth at nature's pace is not simply competitive—it sets the bar. The outdoor environment is richer in ways we may not yet be able to calculate: it gives the pigs more exercise, they have regular access to sunlight and natural rhythms, and in the Missouri University trial, outdoor farrowing appeared to give the pigs an overall lifetime of advantages.

The study confirms similar results and near-countless on-farm experiences of real farmers. The outside hog is permitted the obvious—to be a hog. Exercise builds muscle, the frame grows in response to everyday activity, and the stimulus of a changing environment reduces stress and such environmentally-induced problems as tail biting and other acts of unnatural savagery.

The confinement operator experiences that environment largely from above and over rather short periods of time. He or she is thus above much of the stench, well away from any potential chilling drafts and other discomforts. Confinement workers may also value jump suits and automated auger feeding and choring perhaps too highly. There is an old adage that holds that job one in most confinement units is operator comfort, which in some indoor units now requires that you enter wearing self-contained breathing equipment.

It is painfully obvious that the modern confinement hog is produced almost solely for the greater good of the bottom line of the confinement corporation. The money flows up from the retail counter, but not back down again to the actual raiser. Best estimates now suggest that the really big operations control at least 60% of all butcher hog production. The land-grant, ivory tower thinkers are willing to concede to them a full 80% of total swine production and still believe that all will remain well in the land of the free and the home of the "riblet."

That the confinement producers have the numbers on their side right now, there is no doubt. That they are doing a good job with those numbers is growing increasingly doubtful. Could it be that all that the intensive confinement-rearing system really does is enable fewer people to do a poorer job with more animals?

The Numbers Thing

The numbers question has, on one level or another, been one that the hog industry and individual producers have been wrestling with for nearly the whole of my lifetime.

When my father brought us back to the farm in 1959, 25 sows was a big herd in Missouri and most other places. And yet on my school bus ride home each day, corn or hogs were never out of my line of sight unless we were in town.

Ten sows—a one-boar herd—were a common part of nearly every farming mix. A lot of pigs were farrowed in refitted horse stalls and out in the hickory woods. They were farrowed in spring and fall, and the pigs were marketed through sale barns and into the yards at National City, Illinois. You kept hogs because that was a part of what farming was about.

Let me say upfront that this was no swine-raising ideal, however: hogs were kept on a back burner; many boars were little more than sale-barn sow fresheners; marketing was pretty much dump-and-go; prices cycled often and brutally; and this was a venture a lot of fellows were always saying they wanted out of. The "Big I's"—Illinois, Indiana, and Iowa—held nearly total sway over all things porcine, and whatever direction the business had came from there.

In the mid '60s, the 30 year run of the "hog men" began. People got busy, and particularly more businesslike, when it came to hogs. From Smidley houses to 10-crate modulars to gestation crates stacked two-high, the production technology evolved and boomed. Swine types changed faster than hem lines, and the ideal sow herd ballooned from 50 to 125 to the target production of at least 50,000 butcher hogs a year to be considered a modern mover and shaker.

The belief was that bigger was better, but no one could really explain how being bigger really made things better. The argument of economies of scale never did bear out, and the figures were always being bumped upward. It is now believed that savings of scale don't kick in until about the 3,000-sow level. And that's a mighty big jump from the ideal of 125 sows, the full-time job for one family, that held up for much of my lifetime.

As range producers reclaim a position in farming, they too will have to address the numbers question. This is not a system of production that lends itself to epic numbers, and for at least the immediate future, it won't find markets clamoring for producers with semi-load capacities.

The swine production specialists, even in the go-go '70s and '80s, were always walking too closely to the cliff. When hog prices tanked, very soon whole farms did the same.

There is no ideal number of sows to keep or butchers to raise. You must be guided by what your market will bear at truly profitable levels, and not merely what you can dump onto it. Years ago, I saw a cartoon of Dennis the Menace, which had him and a little friend standing behind a sidewalk lemonade stand with a big sign that read "$100 Per Glass." The caption, addressed to a bewildered-looking buyer, was "Yes, but we only have to sell one glass."

Essentially, the same is true with hogs. Sell them for enough of a profit, and you don't have to sell that many. Go much above 50 sows, and the swine aspect begins to dominate the entire farming operation. Get up around 100 sows, and it becomes full-time employment for at least one person.

I have a friend in Texas producing grass-based Hereford breeding stock, free-range eggs and chicken breeding stock, and he has been exploring range-reared pork as another product for his egg buyers. People buying eggs by the dozen aren't going to buy pork by the truckload.

Three sows farrowing twice a year and producing eight pigs per litter would yield four dozen 240 pound butchers a year. This amount of hog on the hoof could be broken down to yield approximately 7,000 pounds of whole-hog sausage. At $2 per pound, the producer would gross close to $14,000 from a herd of just three sows. Bump that figure to $2.50 a pound, and that gross income figure goes to over $17,000 from a herd of the same three sows.

The numbers to manage here are not number of head raised, but rather selling price that is improved by processing and presentation. Those of us in swine production have yet to grasp the simple truth that the numbers we can produce and those we can produce profitably can often be nearly polar opposites.

To be sure, the organized marketers selling natural pork on a large scale are currently unable to buy all of the animals they need. This is true even with a 5¢ to 10¢ per pound premium on desired animals in place. That breaks down to nearly a 30% premium over some market prices, but even this market will become saturated. Alas, even some of these natural pork products are starting to dabble in big corporation practices.

At a farm show in the early fall of 2003, I was approached by a representative of one of these firms marketing a more natural pork bred from some of the heirloom breeds. He lit me up like a Christmas tree with his talk until he dropped the bomb, which was that he wanted to work only with producers who could deliver up butcher hogs in trailerload lots. Pork is no longer special when it starts coming off of an assembly line. Premium markets are, by definition, rather small and fairly exacting. These markets will reach saturation even faster than the general pork market. And what makes this pork so attractive to its consuming public is its custom-made or boutique-like qualities.

"Boutique" is actually an apt expression here. Pork and nearly any other ag commodity can be produced in the Third World like so many sneakers (and with just about as much flavor). My grandfather's watch is nearly 75 years old, and it's a gem. It continues to keep good time and has actually gained in value, since most people appreciate craftsmanship, especially if it is turned out in modest numbers. A $4 watch from Wal-Mart breaks relatively quickly, is thrown away, and is soon forgotten. To assure ourselves a lasting place upon the land, shouldn't we be focused on creating enduring products that naturally have greater value? Remember that the one great control that still remains with the independent farmer is the flow of production from his or her farm.

A very long time ago, Mahatma Gandhi set down a simple truth that should become Law One in agricultural economics. He said, "The Earth provides enough for every man's need, but not every man's greed." Under Dr. Booker T. Whatley's plan in which even the smallest of farms could earn $25,000 a year, his iron rule was to grow only as much of anything as you could sell quickly and directly. The range producer should then be guided by two simple marketing axioms: (1) make it good, and (2) make it gone.

Could a range operation crank out the 12,500 butchers needed each quarter to make that producer a player, according to current standards? I suspect so. Could it be done at a consistently high profit? Even a continuing profit? That I doubt highly.

Outdoor production is not the way back to 125-sow herds, two big John Deere tractors in every barnyard, and the new Golden Age of the "hog man." It will, however, get hog production back into the hands of independent, fully diversified producers who farm with a different set of numbers and values.

The Modern Pork Maverick

An old joke that has now come to reveal some genuine truths is, "If I had known I was going to live this long, I would have taken better care of myself when I was younger."

Range production in this incarnation is still a relatively young pursuit, and there has to be some concern about how it is tended and operated. Inspiration and information now come at us almost as if they were being fired from machine guns. Evenings find Phyllis and I both reading extensively by the light of the TV, and the bedside radio remains on all night long.

The Internet is expanding as a source of information on alternative ideas and what has come to be considered "nontraditional" agricultural production. I believe it also holds a great deal of potential as a marketing tool. I am part of a group of small farmers that recently sought grants to explore that very concept. Still, we should use what the Internet offers with a grain of salt and a healthy dash of skepticism. A lot that exists there and would appear foolproof simply proves that fools still exist.

A lot of useful information these days will come from some different and, at first glance, quite unusual sources. For example, the many lifestyle publications now available can give you a good feel for the consuming public's mindset and what it is after. They can be very useful in that all-important marketing task of trend spotting.

I now find quite a bit of useful information in sources as diverse as the weekend *Wall Street Journal,* Martha Stewart's writings and TV work, the major news magazines like *Time,* and even *National Geographic.* Sadly, the major farm magazines have become rather disconnected. I have recently seen issues of the three biggest journals that didn't contain a single livestock-related article.

You now have to step outside of the awfully narrow focus that has developed for production agriculture and its minions in recent years. If you want to see just how badly agriculture's big picture has narrowed, just scan a few old farm magazines from the '30s, '40s, and even the '50s.

The range producer must now learn to assimilate a lot of material that appears negative on the surface and still retain a good perspective and form a vision for the future. Hogs have not gone away as a venture for the family farm. They have not been gobbled up by the corporate maw, never to return again. Just look what is happening now with range broilers and eggs from mobile-housed hens.

History is on the side of outdoor swine producers, who have a place alongside producers of grass-fed beef, organic growers, range-broiler raisers, regional produce growers, and many more in that vast quilt of true American agriculture.

Swine production has at times served as a scapegoat for all of the ills of modern production agriculture. Hog raisers have never had the rustic cowboy image to fall back on. This was a pursuit that drew a lot of folks into farming when about all they had to offer was sweat equity. Certainly, there is the mud and muck factor, but have we forgotten what the poet said long ago about the dignity of the plowman on his knees? Since I was a little boy, I have wanted to be a farmer—even a dirt farmer.

The hard going of the mid '90s reinforced the "poor relations" image of pork producers, but we have had a place at the table of American farmers as long as anyone can remember. Farmers were keeping hogs in the Americas long before the Pilgrims got off the boat.

Right now, there is also a real lack of "go to" guys for the outdoor producers. Extension agents were often trained solely in university-owned confinement units. Feed companies now have feed lines targeted exclusively to confinement hogs. And supplies and health care needs have become increasingly difficult to come by in some areas. The good in this is that I haven't had to fend off a feed salesman in a very long time, but the bad part is that I once sold hog feed.

The modern outdoor producer is the new kid on the block, but fortunately it is a very old block that's steeped in tradition. I have found considerable material in vintage ag texts and have an old, dog-eared copy of Morrisons "Feeds and Feeding" that is almost like an old friend. What's more, we're fortunate that there are many veteran swine producers still in place who have a lifetime of experience behind them and who can share their knowledge with the range producers of today.

While I don't mean for this to come across sounding like a paean for positive thinking, the swine business is struggling to work itself out from a hard place. Many producers feel that they have been failed by a calling into which they invested heavily, perhaps too heavily. Range production will keep many diversified farmers in place and producing to a profitable end. What it will not do is bear the weight of the greed and inflated egos that were so much a part of being a "hog man" in the recent past.

Into the Fray

Outdoor production will have many folks entering into what for them is most assuredly a brave new world. With early, small steps, the producer must begin exploring very new and very different options and venues. Trying to find a non-medicated pig starter could take you far, for example. A trip to find a Tamworth boar can take you three states away from home. And getting that first pound of pork sausage sold can be scarier than the whip at the county fair.

We sold our first whole-hog sausage at our local farmers market for $1.75 a pound, or 3 pounds for $5. We labored long and hard over those early figures and struggled with the small farmer's doubts of worth and value. Who did I think I was? Jimmy Dean?

And you know what? To succeed, you are going to have to be Jimmy Dean, or at least the Jimmy Dean for your neck of the woods.

The traditional approach to swine production has been to make a start of it and then just grow and grow like Topsy. The boys at the buying station were there 5 days a week, ready and waiting. Sell enough to them often enough, and surely you will even out all of the highs and lows of the marketing year. When manufacturers can't sell for a profit, they won't produce, and many firms now only build to order to be assured of a profit.

A local producer who works away from home is selling butcher hogs to other workers at his place of employment. He places hogs on feed only when he has a commitment for their purchase at a good price and even requires an upfront $25 deposit to get things rolling. It is a common practice in other markets, and consumers seem willing to accept it in the food sector as well. It is the basic structure of produce growers operating under a structure that has come to be known as community-supported agriculture, or CSA.

The successful direct marketer fosters the impression of openness and accessibility. Niche markets will now often welcome you if you are correctly positioned. Confinement production is now pretty much a closed loop, with animals moving through a series of contract producers and straight on to slaughter. It's odd that often now, where hogs are found in the greatest numbers, they can be the hardest to acquire. One of the reasons for the high prices being paid for show pigs lately is that feeder pigs are simply no longer available, even deep in the heart of the Corn/Hog Belt.

The independent range producer may now be operating the only game in town for many potential hog and pork buyers. The range producer also retains a high level of visibility. Of course, the confinement producers are doing things behind closed doors, since it's not a good image and may actually foster even more consumer concerns about their production practices. I think that one of the reasons small family farmers have such a positive image with consumers is that they have had to do all of their farming right out there in the open.

The front spread or road pasture is a selling tool that has been used by cattle raisers for generations. Most of us have seen beef cows advertised as being "the front pasture kind." Actually, hogs show well on grass, too. White hogs especially seem to show best on a green background. We've had several potential buyers stop by over the years because they'd glimpsed the hogs from the road and had gotten a better look at them in the sale pens Dad had us build along the lane leading to the house.

The range producer now is solidly out-of-doors and out there in the marketplace. As a producer, you will be honing production methods, positioning your product for maximum profit, building a new infrastructure, and creating an important role and identity for yourself.

Change, Change, Change

Throughout my years in the farming life, this calling has at various times been termed "farming," "agriculture," "agribusiness," and now "food science." The Frankenfood people are now using production contracts to relegate grass producers to the status of lab assistant Igors in their confinement labs.

I probably made more real money and was certainly happier back when we were all still real farmers, but change, as they say, is inevitable. Sometimes in swine production, change will sweep through in a big way. It may be matters of swine type this year, or else it could have to do with changes in technology the next. Only in confinement rearing, however, have we seen efforts to rebuild the hog to fit the parameters of the technology. And they're coming up with an interesting specimen. It is certainly not a breed unto itself, but we may be getting fairly close to a Landrace type of something that can be termed a "confinement hog."

They are largely white, quite long, have a thin rind, and have been bred for an exceptionally docile temperament. Mating performance might be considered compromised by extensive use of artificial insemination, estrus management, and life within an extremely Spartan environment. It is clear that (with apologies to the late, lamented Oldsmobile) this isn't your father's hog.

Pork production is now being posited as little more than an extrusion process: a small amount of rather stripped-down swine genetics is poured into one end, and a pork product that would aspire to a consistency of some form of vanilla pudding comes squirting out of the other end. It is "meat" produced by technicians. The question is, is it meat production that has a long-term future? Corporate thinking is to take all of your profit in the here and now.

The drive is for short-term profit, and some confinement units going up now are doing so with a plan of operation of 10 years or less. Contracts are generally written for just 1 year at a time, and there have been several reports of baby pigs being euthanized during times of very depressed prices. This is hog farming and agriculture that's strictly run as a business, nothing more.

All efforts at the moment would seem to be for taking food production to a global stage. Let it flow from the cheapest points to produce and on to the points of greatest demand. If the Third World will let them get away with using banned chemicals and dubious practices, so much the better. How the $2 a day or less most of the world is working for will finance and reward this plan evidently has yet to be determined.

Nor have those economic linchpins of demand—real profitability and consumer acceptance—been satisfactorily explained away in the above grand scheme. With rapidly growing populations of Muslims and others who for one reason or another will not eat pork, it will never be the meat for the masses in the "global market."

Pork raised for the American market will find the greatest demand and the strongest markets when it is produced the American way. The range producer thus now has a coast-to-coast niche market. There are huge markets to be pioneered in this country.

For example, here in Missouri, considerable effort is being expended to bring farmers' markets into the inner-city areas of St. Louis and Kansas City. The folks there have limited transportation options, and yet they are also not so accessible to the farmer, who would have to travel fairly long distances and through potentially difficult traffic conditions. How then to reach this potentially quite strong market for pork and pork products that are so highly valued by the people there?

Heads, snouts, tails, ears, feet, and chitterlings: all these are highly valued by many people around the United States. And yet in some processing systems, they are now little more than throwaways with the rest of the offal. There are many such underserved markets around the country, and many of these cultures have a tradition of buying direct from the farm.

Valves from the hearts of young pigs have been implanted in human hearts, and the skin of young pigs has been used in burn treatments. Hogs are among the very few animals that researchers can get to voluntarily drink alcohol for research work on alcohol consumption and abuse. I cite these to point up the need for swine producers to think beyond pork as a mere commodity. The only thing that can limit the range producer now is, I honestly believe, lack of imagination about what can be done with his or her production.

The outdoor-produced animal is the ideal of the informed consumer—the consumer with the greatest disposable income. The great problem for producers is that we are standing at a moment of transition. We are on the horns of that fabled Chinese curse, "May you live in interesting times." Still, if we handle it correctly and wisely, we can make that inevitability of change problem work to our advantage.

Every trend and fad in swine production in modern times has resulted in a shakeout in producer numbers. The numbers are now turning against the confinement producer. At this moment, a farm with a set of used confinement facilities, especially one with older lagoons, is sharply discounted at the time of sale. A 10-year-old setup has to be considered badly out of date; it may not be restorable and may represent environmental ills that can't be worked out, even with an entire firm of Philadelphia lawyers. A badly designed or poorly functioning lagoon may now sop up hundreds of thousands of dollars in order to be set right, and all across the Midwest, a great number of confinement buildings are sitting idle, moldering away.

The change to outdoor production will be hard for many folks. By my reckoning, there are some third-generation confinement producers on the land now. The first waste pit in our part of the state went in under a farrowing unit when I was still in grade school, and that unit is still standing, although the years haven't been kind to it. For these producers, any kind of change that has even the slightest appearance of being a step backward will be resisted and resented.

Like all good things, it will have to unfold in its own good time. It is being done with three sows here and 10 feeders there, and its adherents are starting to take some stronger leadership roles. They have, for example, taken a strong role in preserving some of the heirloom and minor swine breeds.

I have a friend in western Illinois who has small groups of Berkshire, Tamworth, and Hereford hogs. In fairly short order, he became president of the state breeders' group for one of the breeds and is busy organizing an heirloom breed show for the Illinois State Fair.

When things got really bad back in the mid '90s, I took heart in a local incident. One of the last three known breeders of Mulefoot hogs lived in the county to the north of us. He is a great old gentleman, and he regularly came to auctions we held in the Amish community near Bowling Green, Missouri.

He raised his distinctive solid-hoofed hogs not far from the Mississippi River, in Pike County, home to Mr. Twain's fictional Finn family. His plan for the hogs was, upon his death, to have them all sent to slaughter. He had come to believe that there was no longer any interest in his old breed.

At about that time, efforts by the American Livestock Breeds Conservancy began to pay off, and interest in his Mulefoots and other old swine breeds began to grow. Our friend got calls about his hogs from people as far away as Vermont. Several new herds of Mulefoot hogs were started. New breeders began displaying their animals at farm shows, and our friend's genetics are being put to use in numbers not seen even when hogs were king in the Midwest, 20 years ago.

With numbers beginning to build among range producers, marketing successes like the Neiman program in Iowa, coupled with a natural interest in new ventures by farmers of all kinds, outdoor production has a bright future. Sure, change is inevitable, but sometimes change is for the better. For those who are adamant that taking hogs back outside is a step backward, I would offer them these rich words from the French: "When you reach an abyss, the only safe step is backward."

On Being an Outdoor Producer — Now!

Progress in range production can never be measured simply in terms of pounds of output. The successful outdoor producer will increase income by producing better, more diversified product from the numbers at hand.

There's an old story passed around in our business about the watermelon grower who was told by his tax preparer that he was losing 50¢ on every watermelon he hauled to town. His solution was to buy a bigger truck.

That's a very old joke, although a real story in a similar vein was related to me by a buying-station manager several years ago. He had called aside a farmer we both knew to advise him that his hogs needed some upgrading. A few weeks later, that producer began construction on a five-figure farrowing house. Shaking his head at that turn of events, the buying-station manager told me all he had thought the farmer should do was to buy a couple better breeding boars.

Farmers are eventually going to have to move past the idea that the only way they can increase income is by increasing gross output. It is a very bad setup for Plan A. If your only solution to the problem of losing $5 per head on the 1,000 animals you sold last year is to produce 2,000 this year, you have some real holes in your business plan—about 2,000 of them, to be exact.

If the classic Plan A for financial success has been to increase numbers, Plan B has been to mercilessly cut costs of production. As a fallback position, it has some merit, but always bear in mind the teaching of our Puritan forebears: moderation in all things.

Actually, cost-cutting is an essential part of outdoor production. Investments in facilities, for example, if not totally barebones, are getting close to it. As long as it's safe and comfortable, it matters not to a hog if it's lying in a $10 hut or a quarter-million-dollar building.

Do your cost-cutting judiciously and never forget that true cost-cutting can sometimes mean spending top dollar. Good hogs, for example, are fed and gone from the farm—turning a profit—rather quickly, while some of those bargain-basement purchases just stay on the farm and eat and eat and eat.

A purchase can be considered a worthwhile investment if it improves the end product, pares costs to produce without dampening quality, adds to timeliness and efficiency, and/or brings in a new, profitable market. All of these can advance an enterprise not by making it bigger, but by making it better. Always bear this in mind that if the world will beat a path to your door simply for a better mousetrap, just imagine what it will do for a better pork chop?

Small and moderate-sized producers have been guilty in times past of being rather lax in the quality department. How it is that being small got you a pass on doing a good job I have never understood. I've seen folks struggle hard to hang on to 15, 100, or some other magic number of sows when their real profits were being made with just the top half or even top third of the herd.

Job one when it comes to getting ahead is to get good. And when you get good, you want to strive—using a word my nephew coined when learning to talk—to get gooder.

Don't be afraid to ask yourself the hard questions. Are you now weaning nine pigs per litter? What are your days to market weight? How are your pasture yields? Are those pastures being fully utilized?

You can get bigger and you can get better, and sometimes you can do both. Sometimes, but not very often. A successful pasture operation may be just three sows or 10 feeder pigs big. If those three sows are purebred Black Polands, Landrace, or Durocs, they can produce everything from whole-hog sausage to good breeding boars. Those 10 feeder pigs can be just another ton of meat product, but in the right hands, they can amount to additive-free natural pork fit for popular new chefs and that whole lace tablecloth crowd.

I do a lot of my farming first with a pencil. I take the numbers, the available data, and then try them to see if they will work on our farm. A lesson I simply failed to see in the early going was in the swine sale catalogs that once filled our mailbox every spring and fall. It concerned performance data, but not in the way that many folks think of such things. It wasn't about growth rates or feed efficiency, but rather something even more basic and essential to a successful swine enterprise.

For starters, a sale catalog might present information on a hundred litters and a score or more of line breedings. The idea of producing animals in huge numbers to achieve even a modicum of success works along the same lines as the reasoning that holds that given enough monkeys and enough typewriters, eventually you will get the works of William Shakespeare.

Well, years ago a catalog listing caught my eye and radically changed my way of thinking about what is and isn't truly profitable swine performance. The sale catalog was for a breed not normally noted for large litter size. However, one litter that was listed actually had over 10 head of pigs that were to be offered in a sale of breeding stock. Five were of sufficient quality as breeding boars to be placed in the top half of the sale offering. This was genetic depth of the best and truest sort, and it set my mind awhirl.

The common thinking was that you would need three to four litters of pigs farrowed before you could achieve these numbers. Thus, I began digging deeper into sale catalogs. Some litters, even very large ones, produced just one or two animals that merited being listed in the catalog for sale. Other litters, sometimes very few, were good-sized and contained a high number of quality individuals. Taking time to trace the pedigrees on those exceptional litters showed a consistent pattern in some lines and on some farms.

Later on, I dropped into a seat at one particular sale of offspring of a fairly notable set of boars. Prices started flat and stayed slow and set me to thumbing through the pages of my catalog. Here was a litter of eight, and then several of three, four, or five farrowed. Littermates to the dams of the larger litters, when bred to the same boars, were producing small litters.

The most important trait in swine genetics is live young. If you take nothing else away from this book than that simple truth, you will still be well served.

My studies showed that not only were the higher-yielding litters fairly closely related, but that these impressive figures were being generated in the smaller herds and on the smaller farms.

In livestock breeding, good things don't just happen. There is no serendipity in a sow lot.

Every good swine venture is the unique creation of the producer. It reflects everything from the soil of the home farm to the clarity of the producer's vision. The successful range producer will find that he or she now has more in common with makers of fine wines than with producers of mere agricultural commodities. Range producers must create herds and pork products that are every bit as distinctive as a fine vintage. Range production can be said to be truly custom-designed cottage agriculture. It is unique to the local environment, applies genetics for very specific means and ends, and it gets the farmer involved all the way from breeding pen to the retail process.

The range producer now prevails with a product that transcends the simple generic pork product. This creation must still be considered a work in progress, however, and it's still striving for its fullest recognition.

It is the best production possible, but only for as long as we believe in it and give to it our best efforts. The only Number 4 butcher hogs I ever saw came out of an outdoor operation down along the Missouri River. They were dumped a half-dozen times a year into a local sale barn. The whole thing that led up to this was "dump agriculture": dump the sows out, dump an old boar in, dump together the cheapest feedstuffs possible, dump them into the hogs that lived to eat them, and then dump those that remained for whatever that one localized market would give for them. And that market, any market, "gives" very, very little.

The range producer functions with a very clear vision of the market (or markets) he or she is producing for, the animals the market wants, and the farm layout needed to produce those hogs. Up to the Canadian border, hogs fare quite well out-of-doors for most months of the year. For at least the first half of the 20th century, all hogs were outside virtually the year-round.

In the 21st century, methodology, technology, and demand are coming together to again favor this historically proven method of swine production. The producers at the center of this confluence are being presented with a great opportunity. Thinking of it now, I am reminded of that vow we so often repeated years ago in our high school FFA meetings (it was still the Future Farmers of America then).

The President would ask, "Why are we here?" And the very first words of our chorused response were, "To practice brotherhood, honor rural opportunities . . ."

Outdoor swine production is a great opportunity for the family today. It may be one of the last that we will have that will enable us to keep swine production on the nation's independent family farms. It is an opportunity that must be honored both in principle and in practice.

Looking Forward & Back

Growing up in the Midwest, corn was seldom out of my line of sight. Hogs were as commonplace as the air we would breathe and water we'd drink. I have often gone to sleep to the sound of flopping feeder lids and have risen in cold winter nights to check on the farrowing progress of a gilt or two. I found great inner contentment when our home freezer was filled with pork produced "our" way. This is just the way things had been in our part of the world for many long seasons. Before Mr. Deere's plow and Mr. McCormick's reaper came along, there were American farmers with hogs—a great many American farmers with hogs.

Pasture production is not an aberration. It is, in point of fact, the historical norm. This should be the yardstick by which everything else is measured. Pasture production always worked, and as best I can determine, it was never found wanting.

This is the first way—the only proven, lasting way—to best serve and protect the land, the animals, and the people. Through no fault of its own, range production was largely abandoned as a means of farming for roughly the last 40 years. Four-tenths of a century is short reckoning by any sort of historical or economic standard. Who knows—the slat and crate setup may yet prove to be just one more swine production fad.

Range-raised animals, on the other hand, always have and always will continue to grow. American expansion was very often fed and fueled with cured bacon and hams tied to saddle horns. Pork was often the first product harvested from a homestead and was sometimes so dependable that they helped many a farmer keep the family farm.

Our first old Hamp-cross sow bought back in the early '60s began her life and raised her first couple of litters on a sunny, windswept hillside in eastern Missouri. She wintered in a lean-to attached to an old horse barn, grazed on a wild mix of grasses and legumes, rooted grains of shelled corn from icy winter mud, took to the woods at the foot of that hill for summer shade, and farrowed in small houses that are little changed in materials and design from those built in the 19th century.

She and tens of thousands of others like her earned their keep in a time of dollar-and-a-dime corn and eighteen-cent hogs. It wasn't pretty, but it was dependable. The hogs and the folks who tended them worked together and were much more than cogs in some corporate whirligig.

Back then, hogs were part of a larger picture. We had hogs, beef cows, and a milking cow. There were sheep and chickens, and overall, it was a very good world for a growing boy and his old shepherd dog. We would load hogs onto Mr. Bob Hardesty's truck on a Sunday evening to be sold at the National Stockyards in Illinois the next morning. Dad would shut the truck's end gate and yell to Mr. Hardesty, "John Clay all the way."

This meant that the butchers, boars, old sows, and everything else we were "sending to town" were to be entrusted to a commission seller for the John Clay firm. In those days, lists of sellers in each day's market were phoned to the local radio stations, and it was quite a feather in your hat to have it reported that your hogs topped the market on any given day.

I often think of those days whenever I hear fears expressed about tracking meat animals with country-of-origin labeling. There were times when we would load up 10 butchers, two old sows, cull ewes and butcher lambs, plus a calf or two on Mr. Hardesty's old straight truck on a Sunday evening. In the Tuesday morning mail would be a check and a detailed account of every animal's weight, grade, price per pound, and buyer. And all of this in a time before computers and voicemail. The commission houses did this for quite modest fees and worked hard in our behalf, since their income hinged on our own.

We walked a lot of cheap corn crops to town on four strong legs. We farrowed in old horse stalls and on straw in modified A-frame houses that were of a design that had been in use since before World War I. Dad and I were in place with a few good sows when hog farming was ramped up to swine production and then to a bona fide pork industry. The 30 years from '65 to '95 were a hoot and a half.

I grew up in the eye of the storm, if you will. Some of the legends among purebred hog men were our friends and neighbors. Feeder pigs moved through Missouri by the tens of thousands, headed for the corn of Iowa and Illinois. Confinement-rearing methods were tried and tested in many forms all around us.

What has stayed with me throughout this boom and the many changes we've seen was how the purebred folks, the genetic movers and shakers, remained close to their roots in very many instances. They fiercely clung to the belief that sows must spend at least a major portion of each year out-of-doors. They liked to work those high-dollar boars in dirt lots that had some natural "give" to them.

Young stock might have been grown out on feeding floors with a concrete base, but most of those were at least open to the Sun and the elements. One or two bays might have lowered ceilings and supplemental heat for the very young shoats, but those feeders and waterers were out there where the Sun shined and the snow fell. And well before hitting the sales and the breeding pens, most were moved off of the concrete and onto "dirt" to assure soundness at the time of sale.

In point of fact, "raised on dirt" is a term that's still used to emphasize merit in both breeding animals and meat hogs. The guy or gal farming on dirt actually remained a player for nearly all of those booming, churning 30 years cited above. They may have gotten muddy and wet, but they did something very special in their time upon the land.

I can remember one year when the Missouri State Fair's top Yorkshire boar came off a postage-stamp-sized farm. It was born one cold night in a secondhand Smidley house and grew up in dirt lots stitched together with battered panels and hot wire. When he sold for four figures (pickup truck money then) and went on to take up a prominent role in that breed, it was a shared victory for all of us in knee-boots and frayed denim.

Those folks have never really gone away (we've gotten a little grayer and a lot slower, but we're still around). American agriculture is slowly making its way back home, and they are bringing the "bacon" back home with them. At a recent farmers' market, a Mulefoot gilt found a new home, and three Hereford feeder pigs were sold to a young man that sells butcher hogs and halves to those who work with him in a nearby city. A fellow who lives on the south end of our county and works with recent immigrants from Eastern Europe called me just the other day, seeking hogs for those people.

It is a new day for some old folks and some young ones, too. In March of 2004, over 150 4-H and FFA youngsters weighed in over 400 head of pigs for the market-hog competition at our Lincoln County Fair. The interest in these special animals never goes away.

American agriculture takes its cues from some rather strange places. Many trends in machinery design and livestock keeping begin in Europe and Great Britain. Consumer demand moves inland from the coasts, and trends in agricultural production will more often be patterned after activities in states like Pennsylvania and California than Iowa and Kansas.

More and more, the restaurant trade is demanding and beginning to echo the concerns of its customers, particularly in areas of humane treatment of animals and limited antibiotic use. In Europe the crates are coming out and the barn doors are being thrown open. And even in the big cities that are supposedly best served by large-scale confinement operations, folks are paying $4 a dozen for mobile-housed eggs and over $2 a pound for range broilers. In hog country, there is something in the wind other than the stink of hog-house lagoons, and that something is change.

The range producer, the only one in position for such change, doesn't fit into any standard mold—never has and never will. This is not farming by the seat of one's pants, but it does rely upon a lot of on-the-spot creativity.

In the end, it will make you a farmer of grasses and legumes, a true herdsman, a tinkerer and builder, a watcher of the skies and the seasons, and a visionary in muddy boots. It certainly creates beautiful images of Red and White and Spotted hogs on rolling green pastures, but it is neither magical nor ever unfailing.

Summer pastures will disappear under winter snows, there will be at least two mud seasons a year (we once had a January thaw eat a hog house), hogs will die, prices will fall, and you will catch a bad cold or flu during most farrowing seasons. Cattle producers now are working on grazing patterns that will keep their animals on pasture for up to 10 months of each year, well into the upper Midwest. Hog raisers aren't that fortunate, as their animal of choice is the one that made the Corn Belt a reality. From Cincinnati to Chicago to Kansas City, America's first great wave of wealth included such names as Armour and Swift, the packers and princes of pork.

In the most accurate sense, the range producer is a pasture/wooded lot/corn field/drylot producer. Your hogs will not exactly be constantly in motion, but you will soon see the value of a couple of good Blue or Red Heelers or some Border Collies. One of Missouri's noted Hampshire breeders started many of his swine sales by offering a couple of "started" herding dogs and a few puppies. The started dogs often rivaled some of his best-selling boars in price, too.

Outdoor rearing is the centuries-old norm for swine production. We know of swineherds from as far back as Biblical times. And yet range production now is not simply the old made new again.

It is still tough work sometimes. There will come cold January mornings when you will think that they can't give you enough free feed-company caps to continue doing this, and you will be right. The perks in the swine-farming life can be few and far between. For this reason, you will have to earn your own perks by making sure you have an explicit goal of producing optimum returns from the numbers you can comfortably handle on your farm.

Success at all costs won't cut it here either, since the producer's best rationale for more natural production is black ink at the bottom of the ledger. And here the range producer can take on and beat the confinement sector at the most important aspect of their game. Besides, make enough money and you can buy your own caps, ones that say whatever you want them to say.

A new era of livestock production is just now beginning. It is coming at a problematic time as the American farm population is rapidly graying and the average farmer now is crowding 60. I know this because I am a member of that generation.

Changes come hard at this age, but come they must, for the simple reason that so many of us are getting older. If the family farming tradition is to survive, there will have to be more ways into production agriculture than to marry or buy your way onto a farm.

Land at four and five figures an acre means fewer and fewer big farms being launched, and also that a great many smaller agrarian parcels are emerging. These new small farmers are big players, buying the lion's share of new tractors (most under 60 horsepower), producing a great many specialty crops, and being key to the revival of everything from organic production to fine wines to fancy table eggs.

This new blood, with their values and their goals, will shape a good portion of swine production in the future, just like we're seeing for poultry production now. Range pork is now at the point where the range broiler was 20 years ago and grass-fed beef was 15 years back. It is coming, you can be assured of that.

Once again here in Missouri, there are young people with 40 hour jobs and a beginner's handful of sows. Confinement units are standing empty above pits and lagoons that bubble and gurgle ominously.

And the fellows my age still read the few remaining hog magazines, meet and talk in little knots at the feed stores and the farm supplies shops, and they have just enough left to give it one more shot when the time is right again.

I've had hogs in good times and bad. I've worn out Lord only knows how many pairs of knee-boots, been flipped over a gate or two, topped a couple of markets, and have produced some good ones along the way. I've done chores on cold, rainy mornings and late into warm, soft spring evenings.

At the end of such days, when I would leave the hogs in warm, dry straw beds or in good grass as the first stars were appearing, I always felt good about them and myself. I feel it was where they would have chosen to be, in a role and a place natural to them, and they were both contented and productive.

It is a full, harmonious life that's enough for them and for me. Can anyone ask more from a calling?

INDEX

Also from Acres U.S.A.

Reproduction & Animal Health

BY CHARLES WALTERS & GEARLD FRY

This book represents the combined experience and wisdom of two leaders in sustainable cattle production. Gearld Fry offers a lifetime of practical experience seasoned by study and observation. Charles Walters draws on his own observations as well as interviews with thousands of eco-farmers and consultants over the past four decades. The result is an insightful book that is practical in the extreme, yet eminently readable. In this book you will learn: how to "read" an animal, what linear measurement is, why linear measurement selects ideal breeding stock, the nuances of bull fertility, the strengths of classic cattle breeds, the role of pastures, the mineral diet's role in health. *Softcover, 222 pages. ISBN 0-911311-76-9*

Homeopathy for the Herd

BY C. EDGAR SHAEFFER, V.M.D.

Subtitled *A Farmer's Guide to Low-Cost, Non-Toxic Veterinary Cattle Care,* this new information-packed book by *Acres U.S.A.'s* Natural Vet will tell you what you need to know to get started in the use of homeopathic medicines with cows. Using case studies and practical examples from both dairy and beef operations, Dr. Shaeffer covers such topics as: creating a holistic operation; organics and homeopathy; prescribing; mastitis and fertility-related problems; and the *Materia Medica,* keynotes and nosodes. Also includes a convenient section that lists specific conditions and remedies. *Softcover, 222 pages. ISBN 0-911311-72-6*

Herd Bull Fertility

BY JAMES E. DRAYSON

James Drayson spent a lifetime researching and teaching about cattle breeding and fertility, with 35 years of experience in measuring bulls from a fertility standpoint. He followed 1,500 bulls from birth to death and recorded findings that are unequaled by any other researcher. *Herd Bull Fertility* will teach you how to recognize whether a bull is fertile even before the semen test. This manual, generously illustrated with photographs and diagrams, is a must for the cattle grower choosing a bull for his breeding program. *Softcover, 135 pages. ISBN 0-911311-73-4*

Natural Cattle Care

BY PAT COLEBY

Natural Cattle Care encompasses every facet of farm management, from the mineral components of the soils cattle graze over, to issues of fencing, shelter and feed regimens. *Natural Cattle Care* is a comprehensive analysis of farming techniques that keep the health of the animal in mind. Pat Coleby brings a wealth of animal husbandry experience to bear in this analysis of many serious problems of contemporary farming practices, focusing in particular on how poor soils lead to mineral-deficient plants and ailing farm animals. Coleby provides system-level solutions and specific remedies for optimizing cattle health and productivity. *Softcover, 198 pages. ISBN 0-911311-68-8*

Natural Goat Care

BY PAT COLEBY

Goats thrive on fully organic natural care. As natural browsers, they have higher mineral requirements than other domestic animals, so diet is a critical element to maintaining optimal livestock health. In *Natural Goat Care,* consultant Pat Coleby shows how to solve health problems both with natural herbs and medicines and the ultimate cure, bringing the soil into healthy balance. Topics include: correct housing and farming methods; choosing the right livestock; diagnosing health problems; nutritional requirements and feeding practices; vitamins and herbal, homeopathic and natural remedies; psychological needs of goats; breeds and breeding techniques. *Softcover, 374 pages. ISBN 0-911311-66-1*

Natural Horse Care

BY PAT COLEBY

Proper horse care begins with good nutrition practices. Chances are, if a horse needs medical attention, the causes can be traced to poor feeding practices, nutrient-deficient feed, bad farming and, ultimately, imbalanced, demineralized soil. Pat Coleby shares decades of experience working with a variety of horses. She explains how conventional farming and husbandry practices compromise livestock health, resulting in problems that standard veterinary techniques can't properly address. *Natural Horse Care* addresses a broad spectrum of comprehensive health care, detailing dozens of horse ailments, discussing their origins, and offering proven, natural treatments. *Softcover, 164 pages. ISBN 0-911311-65-3*

To order call 1-800-355-5313
or order online at www.acresusa.com

Eco-Farm: An Acres U.S.A. Primer

BY CHARLES WALTERS

In this book, eco-agriculture is explained — from the tiniest molecular building blocks to managing the soil — in terminology that not only makes the subject easy to learn, but vibrantly alive. Sections on NP&K, cation exchange capacity, composting, Brix, soil life, and more! *Eco-Farm* truly delivers a complete education in soils, crops, and weed and insect control. This should be the first book read by everyone beginning in eco-agriculture . . . and the most shop-worn book on the shelf of the most experienced. *Softcover, 476 pages. ISBN 0-911311-74-2*

Weeds: Control Without Poisons

BY CHARLES WALTERS

For a thorough understanding of the conditions that produce certain weeds, you simply can't find a better source than this one — certainly not one as entertaining, as full of anecdotes and homespun common sense. It contains a lifetime of collected wisdom that teaches us how to understand and thereby control the growth of countless weed species, as well as why there is an absolute necessity for a more holistic, eco-centered perspective in agriculture today. Contains specifics on a hundred weeds, why they grow, what soil conditions spur them on or stop them, what they say about your soil, and how to control them without the obscene presence of poisons, all cross-referenced by scientific and various common names, and a new pictorial glossary. *Softcover, 352 pages. ISBN 0-911311-58-0*

The Biological Farmer

A Complete Guide to the Sustainable & Profitable Biological System of Farming

BY GARY F. ZIMMER

Biological farmers work with nature, feeding soil life, balancing soil minerals, and tilling soils with a purpose. The methods they apply involve a unique system of beliefs, observations and guidelines that result in increased production and profit. This practical how-to guide elucidates their methods and will help you make farming fun and profitable. *The Biological Farmer* is the farming consultant's bible. It schools the interested grower in methods of maintaining a balanced, healthy soil that promises greater productivity at lower costs, and it covers some of the pitfalls of conventional farming practices. Zimmer knows how to make responsible farming work. His extensive knowledge of biological farming and consulting experience come through in this complete, practical guide to making farming fun and profitable. *Softcover, 352 pages. ISBN 0-911311-62-9*